A SEARCH
FOR
ORIGINS

A SEARCH FOR ORIGINS

SCIENCE, HISTORY AND SOUTH AFRICA'S 'CRADLE OF HUMANKIND'

Edited by
Philip Bonner
Amanda Esterhuysen
Trefor Jenkins

WITS UNIVERSITY PRESS

Published in South Africa by:

Wits University Press

1 Jan Smuts Avenue

Johannesburg

2001

http://witspress.wits.ac.za

Entire publication © Wits University Press 2007

Foreword, introductions and chapters © individual authors 2007

First published 2007

ISBN 978-1-86814-418-1

Edited by *Karen Press*

Picture edit by *Sally Gaule*

Layout and design by *Abdul Amien*, Cape Town, South Africa

Printed and bound by *Paarl Print*, Paarl, South Africa

Contents

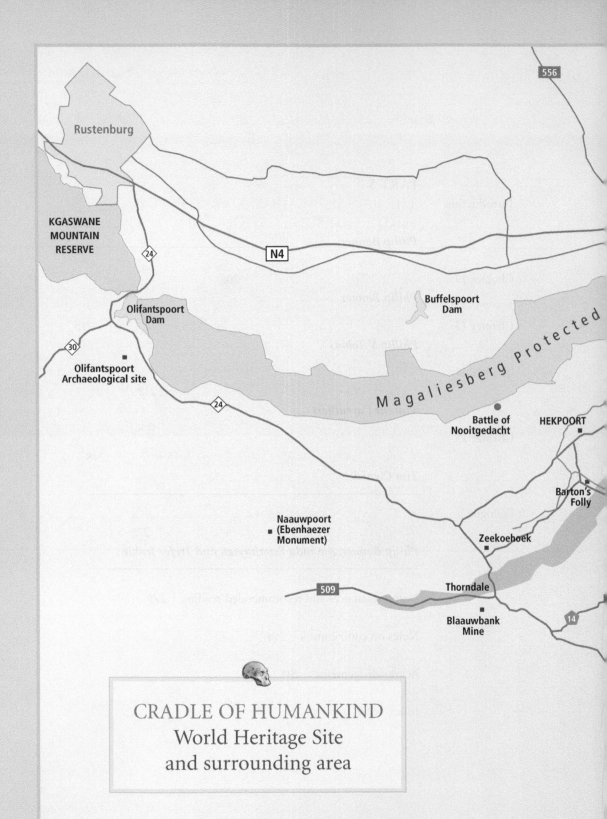

Rustenburg

556

KGASWANE
MOUNTAIN
RESERVE

24

N4

Olifantspoort
Dam

Buffelspoort
Dam

30

Olifantspoort
Archaeological site

24

Magaliesberg Protected

Battle of
Nooitgedacht

HEKPOORT

Barton's
Folly

Naauwpoort
(Ebenhaezer
Monument)

Zeekoehoek

509

Thorndale

14

Blaauwbank
Mine

CRADLE OF HUMANKIND
World Heritage Site
and surrounding area

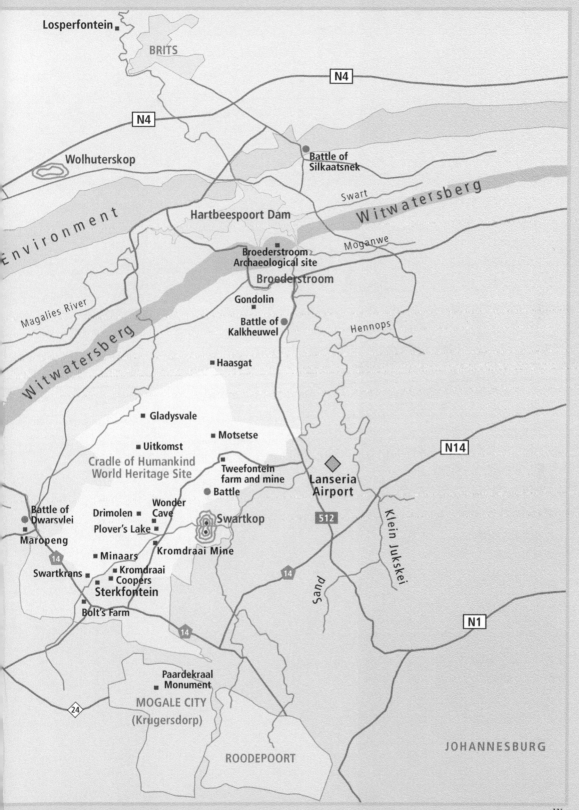

THE DEEP ROOTS OF HISTORY AND PREHISTORY AROUND STERKFONTEIN, SOUTH AFRICA

Phillip V Tobias

The need for this book has arisen from the listing as a World Heritage Site of the *Fossil Hominid Sites of Sterkfontein, Swartkrans, Kromdraai and the Environs* – popularly though not entirely accurately called the 'Cradle of Humankind'! The World Heritage Centre falls under UNESCO, that is, the United Nations Educational, Scientific and Cultural Organisation. So, to understand the roots of this book, it is necessary to explore South Africa's relations with UNESCO and the events leading up to the listing of this country's first World Heritage Site in 1999.

Although South Africa was a founder member of UNESCO in 1946, it saw fit to withdraw from the Organisation in 1956, when the apartheid policies of the Union government were rising to a crescendo of racial intolerance and discrimination. With the perspective of hindsight, there is no doubt that it was UNESCO's vigorous programme against racism that was the cardinal factor in South Africa's decision to withdraw from the Organisation. A resolution adopted at the Fourth General Conference of UNESCO in 1949 called on the director-general, Dr Jaime Torres-Bodet, an esteemed Mexican poet, to (1) collect scientific materials concerning problems of race; (2) give wide diffusion to the scientific information collected; and (3) prepare an educational campaign based on this information. UNESCO in its early years embarked on the preparation of two sets of important publications, *The Race Question in Modern Science* and *The Race Question in Modern Thought*. Collectively these books constituted a scholarly and powerful indictment of racism (or racialism, as it was then still called). They were published less than a decade after the horrors that had been perpetrated during the Second World War in the name of race.

Torres-Bodet convened a Committee of Experts on Race Problems. These individuals were drawn from the fields of physical anthropology, sociology, social psychology and ethnology. Their deliberations led to the 1950 UNESCO Statement on Race (Paris,

July 1950). This First Statement on Race opened with the ringing affirmation: 'Scientists have reached general agreement in recognising that mankind is one: that all men belong to the same species, *Homo sapiens*.' The statement went through several subsequent revisions, the Second Statement appearing in Paris in June 1951; the Third in Moscow in August 1964 (when I was invited by the World Health Organisation to be its representative on the Committee – my own book entitled *The Meaning of Race* having been published in Johannesburg in 1961, to be followed by a revised and enlarged second edition in 1972); and the Fourth Statement produced by a Committee of Experts on Race and Racial Prejudice in Paris in September 1967. Towards the end of the century there was a move for a further revision in the light of newer research, and once more I was involved in these discussions.

The 1950 and 1951 Statements on Race affirmed concepts diametrically opposed to the beliefs and policies of the South African apartheid government, which had come to power in 1948. The latter could not tolerate its citizens being exposed to such 'subversive', 'communistic' and 'liberalistic' information as was emanating from UNESCO. The Union of South Africa's 'complaints against UNESCO' included (1) UNESCO's race policies; (2) various requests by UNESCO to send missions to the Union; (3) 'misrepresentation of the Union's policy' in UNESCO pamphlets; and (4) a 'negligible return' received from the Union's financial contribution. There were protracted negotiations. Then South Africa's ambassador in Paris gave notice on 31 December 1956 of his government's intention to withdraw from the Organisation. This fateful decision took South Africa out into obscurantist and Stygian isolation – and there the country remained for close on forty years.

Among the UNESCO programmes to which South Africa was thereby denied access was the World Heritage Centre and the listing in it of sites, natural and cultural, deemed to be of outstanding universal value. That initiative was adopted by UNESCO in 1972. By the time President Nelson Mandela's democratic government took South Africa back into UNESCO, some 500 sites around the world had been listed, but alas, pitifully few in Africa and none in South Africa.

By 1994 I had been in charge of the excavation of the Sterkfontein cave for close on thirty years. During this time my assistants and I (especially Alun Hughes, David Molepole, Nkwane Molefe, Stephen Motsumi, Isaac Makhele, Hendrik Dingiswayo, Solomon Seshoene and, latterly, Ronald Clarke, Abel Molepole and Lucas Sekowe) had extracted many tens of thousands of vertebrate fossils. These were mainly of mammals, but also of birds, reptiles and amphibians, as well as fossil plants. A remarkable group among the mammalian fossils was that of fossil primates, comprising essentially cercopithecoids (baboons and monkeys) and hominids (members of the family of humankind).

The hominids were especially significant. They fell into two or three species, mostly into *Australopithecus africanus* to which the 'Taung child' belonged. These Sterkfontein hominids extended in time from about 3.3 million years ago to less than 1 million years ago. Most excitingly, with hundreds of individuals represented, palaeoanthropologists for the first time were able to identify males and females, infants, children, adolescents and adults, within the same species. We were able to study fossil populations and not merely single individuals, as was the case when the child skull from Taung came into Raymond Dart's hands in 1924. All told, if we included the Sterkfontein hominids in the Transvaal Museum, which Robert Broom and John Robinson had recovered, and the six hundred in the Wits Medical School's Anatomy Department, which I and my team over the last forty years had brought to light, Sterkfontein had given an unequalled stockpile of over seven hundred fossil hominid specimens from this single cave alone. In its yield of fossil hominid specimens it is the world's individually richest cave site.

Because of this unprecedented plethora of specimens, I took steps in the early 1990s to apply for the Sterkfontein caves to be listed as a World Heritage Site. The government had enacted legislation for the proper procedures and a national committee had come into being under Makgolo Makgolo. A visit by a representative of the World Heritage Centre led to our recognising that listing should not be sought for Sterkfontein alone. As Swartkrans was individually the second richest site and as Kromdraai was historically the first site to yield another type of ape-man, the robust species, and moreover, as there were at least another ten fossil-bearing sites in the vicinity, most of which had yielded hominid remains, the entire area embracing all of these caves – and others that might come to light later – should be considered as a composite 'site', under a single all-encompassing title. After much mapping, describing, photographing and surveying, our application, with a veritable barrow-load of supporting documents, was lodged at the Centre in Paris on 30 June 1998. There followed nearly eighteen months of visits and inspections before the Centre was satisfied that the application was ready to be laid before the World Heritage Committee. When it met at Marrakesh, Morocco, on 1 December 1999, the first three South African nominations were considered: Robben Island, Greater St Lucia Wetlands Park and the Fossil Hominid Sites of Sterkfontein, Swartkrans, Kromdraai and the Environs. All three proposals were adopted unanimously and the sites were listed on 2 December 1999.

Overall the Cradle forms a complex of Late Pliocene and Early to Middle Pleistocene sites. They comprise some of the most important depositories of fossil bones dated between 3.5 and 0.5 million years ago. Their contents have thrown a flood of light on crucial stages in the evolution of humankind. Since then, two virtual outliers of the Cradle have been recognised and listed: they are the historically significant Buxton

Limeworks at Taung, some 500 kilometres south-west of Sterkfontein, and the Makapan Valley ('Makapansgat') some 300 kilometres north of Johannesburg.

According to the new South African legislation each World Heritage Site falls under the charge of the province within which it lies. Most of the Cradle caves are in Gauteng and it has fallen to this province to develop the site. With the enormous enthusiasm of the premier, Mbhazima Shilowa, and his provincial administration, together with the University of the Witwatersrand (which has legal ownership of Sterkfontein and Swartkrans) and a consortium of highly talented planners, Sterkfontein has been transformed into a centre with a museum, restaurant, curio shop, conference centre and exhibition hall, and much improved access to the underground caves themselves. While plans are afoot to attract scores of visitors, tourists and schoolchildren every year, great care has been taken to safeguard the scientific work in the caves and in newly constructed laboratories close by. The planners and the researchers work in close co-operation with SAHRA (the South African Heritage Resources Agency) to ensure that the highest standards are maintained and the accepted practices required by UNESCO adhered to.

Another development is the erection of Maropeng, a new and highly impressive visitor and conference centre about 8 kilometres from Sterkfontein. The word 'Maropeng' is Setswana for 'the place where we once lived'. It comprises a spectacular tumulus within which are a world-class conference centre, exhibition hall and restaurants, with a patio commanding lovely views of a large part of the Cradle and the westerly-flanking Magaliesberg range. Adjacent is a newly opened Maropeng boutique hotel.

Already the dolomitic countryside from almost as far north-east as the Hartbeespoort Dam, for 25 kilometres south-westwards, beyond Swartkrans, Sterkfontein and Bolt's Farm, and westwards to Maropeng, has been transfigured. Myriad opportunities have been created for the economic upliftment of what was a desperately needy area. At Sterkfontein and at Maropeng scores of jobs have sprung forth and continue to burgeon with each new development at the Cradle. This social function has from the beginning been an avowed aim of Gauteng province, the planners and the Wits scientists, supported by property owners in the area.

Excavations are proceeding at the two richest cave sites, Sterkfontein and Swartkrans, and at almost a dozen other sites in the 47 000-hectare Cradle. These 'digs' continue to attract leading scientists, research students and postdoctoral fellows from South Africa and other parts of the continent, and from North America, Europe, Asia and Australia. This book gives an outline of the scientific lure for those who are impassioned by the story of humanity's past. It tells of the progressive unravelling of the tale of discovery over more than eighty years. It reveals that the study of biological evolution is a study of change in animals and plants with time. Little in evolutionary studies, whether based on fossils, molecular analyses or the cultural objects crafted by ever-transforming humankind, makes sense other than through the prism of time. Time helps us to classify the earlier forms of life represented in the fossil record and to tidy up the evolutionary relations among species and genera. Reliable estimates of the dating of fossil remains enable us to relate changes among hominids at any time with the state of stability or change in contemporaneous non-human animal groups. If dramatic changes in a number of animal and plant groups occurred at the same time, such *synchronic* changes need to be explained. They may lead us to recognise catastrophic events, such as mass extinctions, meteoritic impacts, ecological, geological and climatic occurrences. Synchronicity among clusters of events can open our eyes to the causes that may lie behind some evolutionary changes.

Our book summarises some major phases in the history of near-human and human life, against the background of what the English poet Wystan Hugh Auden spoke of as 'Time the Refreshing River'. We can do this today more reliably than before, because of amazing developments in the dating of the past over the fifty years since Libby introduced the carbon-14 dating technique.

This Foreword would be incomplete without my referring to earlier publishing ventures on the 'Cradle'. There were several communications to the South African Geological Society by the Society's co-founder, David Draper, from 1895–1898 following the collection by him of fossils at Kromdraai in 1895. The earliest article to report the recovery of fossil bones within the Sterkfontein caves was published by a party of Marist Brothers from the Sacred Heart College in Koch Street, Johannesburg. Their account of the interior of the caves and of the fossil bed appeared in the French periodical *Cosmos* early in 1889. The author of this account was given as 'Un Frère Mariste de Johannesburg'.

In 1929 the rights to the Sterkfontein caves were acquired by Robert Cooper of Krugersdorp. His quarryman, George W Barlow, had been a quarryman at Taung when the *Australopithecus* child was found in 1924. Cooper and Barlow recovered many fossils from the lime-working activities at Sterkfontein. Good specimens were displayed on a table in a *pondokkie* and these were sold to visitors. Some were pilfered by light-fingered rummagers. The table with specimens was seen by me on my first excursion to Sterkfontein in May 1945. To encourage visitors and promote sales, Cooper produced a small guidebook in 1935, telling the readers about the fossils and prophetically coaxing them with the declamatory invitation: 'Come to Sterkfontein and find the Missing Link'. That is precisely what Broom did in the following year! On 17 August 1936, he received the first hominid specimen from the hands of Barlow. The first scientific publication was that of HBS Cooke, a member of the Geology Department at Witwatersrand University, and one who was destined to have a long and distinguished career as a palaeontologist, specialising particularly in the proboscidean, pig and horse families. His short article, 'The Sterkfontein bone breccia: a geological note', appeared in the *South African Journal of Science* in December 1938. In the early 1950s Professor Lester King of the University of Natal made a preliminary set of comparative studies on the life history of the Plio-Pleistocene caves, to be followed by a compendious doctoral study by CK Brain of all South African hominid-bearing caves known up to that time (1958).

In a letter to me on 10 June 2006 from Basil Cooke, then living in British Columbia, Canada, he reminded me that in 1954 C (Peter) van Riet Lowe had wanted the university or his department (archaeology) to produce a book or booklet on the Sterkfontein caves. A number of people were approached and draft articles with illustrations were collected by Cooke. In his recent letter Cooke reported that the project had died, and he offered to return the chapters. In 1966 the Witwatersrand branch of the South African Archaeological Society asked me to arrange for the preparation of a popular book for sale at Sterkfontein and elsewhere. Sadly that project also fizzled out. A booklet on the Sterkfontein caves by Harry Zeederberg helped to fill the gap in the public's reading. In 1969, HBS Cooke published a comprehensive article on Sterkfontein, in *Studies in Speleology*, under the title 'Preservation of the Sterkfontein Ape-Man cave site, South Africa'. Reprints of that article were on sale at the old Sterkfontein cafeteria for many years. More recently, an up-to-date booklet was published by Ronald J Clarke and Kathleen Kuman in 2000.

I first began to take an interest in the history of Sterkfontein when I was a second-year medical BSc student 62 years ago. At that time I fell under the spell of my anatomy teachers, Professor Raymond Dart, Dr Alexander Galloway and Dr Lawrence Herbert Wells. To these important influences in my early academic life were added the following year Dr Robert Broom, Professor (Peter) van Riet Lowe and Mr BD Malan.

My unpublished report on the science students' excursion (I hesitate to call it an expedition) to Sterkfontein and Kromdraai in April 1945 was my first attempt to set down something about Sterkfontein; it was then only nine years since Broom had recovered the first australopithecine there and seven years after the find at Kromdraai. The discoveries at Makapansgat and Swartkrans were still two and three years respectively in the future.

A few years later, Broom's *Finding the Missing Link* (1950) revealed to me more about this most exciting of subjects and the larger-than-life personalities that pursued it. My own active involvement with Sterkfontein and the Makapan valley in the middle to late 1940s, with Taung in the early 1950s, and with Olduvai and other East African discovery-sites from the late 1950s onwards, cemented my interest in the history of the subject. Then, in 1959, came Berry Malan's unexpected revelations on earlier – nineteenth-century – references to Sterkfontein.

Still later, on the occasion of the Broom centenary in 1966, I lived through and penned further episodes in the history of Sterkfontein. In 1973 my address to the South African Biological Society at the University of Pretoria was published in that Society's journal under the heading 'A New Chapter in the History of the Sterkfontein Early Hominid Site' (1973). During the late 1960s and the 1970s, I published a number of bits and pieces on the Sterkfontein dig and its history, especially on the excavation upon which Alun Hughes and I had embarked in December 1966.

In 1978 the chairperson of the Johannesburg Historical Foundation invited me to deliver the Annual Commemorative Lecture to the Foundation. This spurred me to go more deeply into the history of Sterkfontein. It impelled me to gather together all the references and tidbits I had been accumulating for years on this subject. Although my main interest at Sterkfontein remains its prehistory, this probe into its history proved an exercise of absorbing fascination. My lecture under the heading 'A Hundred Years of History at Sterkfontein' was delivered on 18 September 1979. It had developed into a manuscript of over one hundred pages, excluding the illustrations; and I was persuaded to submit the manuscript for publication as a book to the Witwatersrand University Press, once I had brought the text up to date. In the meantime, at the instigation of one of the editors of this, Professor Philip Bonner, my chapter included here represents an abridged version of my 1979 history, and I am enormously indebted to him for taking on the task of preparing this heavily shortened version.

Phillip V Tobias

University of the Witwatersrand,
Johannesburg

Munster map (1544–45)

PART 1

INTRODUCTION

AFRICA IS SELDOM WHAT IT SEEMS

Philip Bonner

The 'Cradle of Humankind' opens windows onto many pasts: onto the origins and evolution of humanity; onto the ever-expanding frontiers of science in the fields of palaeontology, geology and genetics; onto the remarkable group of scientists that drove these frontiers forward; onto several of the most momentous and formative moments of South Africa's more recent history, including its peopling by its present African population; and onto the often strange, sometimes tortured psyche of a large slice of white South Africa after it came to dominate the sub-continent in the late nineteenth century. The 'Cradle of Humankind' can thus deservedly claim a special status among the heritage sites recording South Africa's past, offering as it does a privileged vantage point from which to understand what it means to be human and what it meant and currently means to be South African.

Part of the enduring appeal of the Cradle is that what it means and what it stands for are in a constant state of motion or flux. The boundaries of science at the Cradle are regularly pushed forward through new fossil discoveries, as can be seen most graphically perhaps in Ron Clarke's unique discovery of a near-complete early hominid skeleton, which has been nicknamed 'Little Foot', and through entirely new domains of science previously unimagined (such as DNA studies). Even though early hominid fossils were recognised at Taung and Sterkfontein in the 1920s and 1930s, their true antiquity and hence significance was only grasped after the Second World War. Similarly, the way in which the Cradle region marks and bears witness to many of the key phases of more recent South African history has only been perceived by scholars in the last thirty years, and has still to filter fully into the wider public consciousness. One purpose of this volume is indeed to help make this knowledge more public.

Partly because of these scholarly and scientific advances, partly because of political shifts, the place of the Cradle in South Africa's public life has also been subject to change. The paradoxes that this raises are noted in Saul Dubow's chapter in this volume. One of the prime patrons of the Cradle's claim to be the site of the origins of humanity was South Africa's four-times prime minister and public intellectual, Jan Christiaan Smuts. It was Smuts who in 1925 gave the Sterkfontein area the name which subsequently stuck – 'The Cradle of Mankind' (though it would later be amended to 'Cradle of Humankind' to reflect a political commitment to inclusivity). However, as Dubow shows, Smuts saw in the Cradle a kind of charter for white domination in South Africa. It was white South African science that first understood the revolutionary significance of the Cradle's fossils, which not only placed local scientists on a par with those in Europe and America, but also attested to a collective white claim to superiority and domination over the majority of South Africa's black population. This is an issue that will be returned to later in this introduction and in the introductions to Parts 4 and 5. From the 1950s on, these assertions of white/European racial distinctiveness and superiority have been challenged by scientists and geneticists in South Africa, notably Phillip Tobias, Trefor Jenkins and Himla Soodyall (all contributors to this volume). Public consciousness lagged behind, but began to catch up in the course of the momentous political transformations of the 1980s and 1990s. Since then the Cradle has become a national icon of a different sort, as a site of reverence and pilgrimage for all of South Africa's – and indeed the world's – citizens, black, brown and white – the place of their beginning. No doubt newer ideas and further changes will unfold. As President Mbeki observed in an evocative phrase, in his address at the opening of the Maropeng Heritage Centre at the Cradle, 'Africa is seldom what it seems.'

The Cradle area cannot be divorced from its immediate neighbour to the west, the Magaliesberg valley and hills, with which it has enjoyed an intimate association. A constant traffic passed between these two areas from the earliest times. For much of its existence the warmer, wetter and more naturally blessed Magaliesberg teemed with game, fauna and life of all sorts. The Cradle, by contrast, was a colder, drier, less well-favoured environment, even in the wetter periods like the Pleistocene in which, as Marion Bamford describes in Chapter 5, lianas and forest shrubs flourished around its rim. In cold periods like the mid-Holocene (6 000–4 000 years ago) it appears to have been abandoned altogether. In the massive sweeps of evolutionary time in between, however, the Cradle/Magaliesberg area constituted a fertile habitat exploited by our hominid ancestors. From the time of the more versatile generalist *Homo ergaster* (see Chapter 6 by Amanda Esterhuysen), hominids must have spent the bulk of their time in the Magaliesberg region, only venturing into the Cradle area intermittently or at specific times of the year.

One feature which clearly distinguished the Cradle from the Magaliesberg was its distinctive geology, which allowed extensive caves to develop among its dolomitic formations and in which large quantities of travertine or limestone built up. During the time of the earliest australopithecine hominids, these caves were the lair of predators like (giant) hyenas and big cats. Our relatively small-framed early hominid ancestors presented a natural prey for these cats, and therefore gave the caves a wide berth, but the remains of those killed and dismembered were often dragged or washed into the caves to be deposited in the lowest hominid fossil-bearing geological members. Later species of hominid who had mastered the use of fire camped on the rims of these caverns, around the protective circle of flames towards which predators dared not venture (see CK [Bob] Brain's discussion of fire, in the interview extract included in Chapter 6). These hominids, too, sometimes fell into the caves, or their remains were washed in, ending up as part of the breccia that was later located there. One of the abiding ironies of the history of this area is thus that the area labelled the Cradle could probably be more appropriately called the Grave.

The Cradle has been the scene of numerous epic battles since the dawning of human time, not only between the australopithecines and the great cats (see Chapter 3 by Kevin Kuykendall), but also between early *Homo* and later australopithecines (see Kathy Kuman as cited in Chapter 6), between many African chiefdoms that settled or tried to settle in the interior and those on the north-moving frontier (in the period sometimes dubbed the *difaqane* – see Chapters 10 and 11 by Simon Hall and Jane Carruthers), between African chiefdoms and Boers in the early stages of the Great Trek (see Chapter 11) and between Boers and Britons in the South African War of 1899–1902 (formerly known as the Anglo-Boer War – see Chapter 14 by Vincent Carruthers). The Cradle thus provides a lens through which to view and comprehend a series of absolutely pivotal and formative moments of South African prehistory and history. Each finds a place in this volume.

The scientific history of the Cradle is peopled by a cast of larger-than-life personalities – Raymond Dart and Robert Broom up to the late 1940s, Phillip Tobias, Bob Brain and Revil Mason thereafter. On the whole, the pre-war personalities were more colourful and/or eccentric. A substantial literature documents their achievements, but even that which indulgently records their foibles tends to cast them in a heroic mould. Yet, no less than Africa generally, these celebrated figures are seldom what they seem. Inevitably their characters were in one way or another flawed; inevitably they had imbibed some of the intellectual ethos and social climate of their time. In pre-war South Africa this inevitably meant that they harboured some sense or other of white superiority which we would now label prejudice. In this they were simply creatures of their time, sharing in the failings of a whole generation, and do not deserve to be individually or excessively blamed. What merits recognition in this context is that they were at least partially able to break through and break free of the intellectual and social straitjacket of their era.

From a scientific point of view, their contributions are both unquestionable and unshakeable: they caused paradigmatic shifts. This was eventually recognised not only by scientists but also in a wider public sphere. Dart, Brain and most recently Lewis-Williams (a student of culturally modern humans), whose writings apply to but are not specifically generated by the Cradle (and whose work is discussed in Chapter 8 by David Pearce), have through the books of popular writers – whose minds were gripped by both the science and the personalities of their subjects – left a huge imprint on the popular imagination: Dart, through Robert Ardrey who wrote the hugely successful and influential *African Genesis*; Brain, through Bruce Chatwin, the doyen of popular travel writers in the 1970s and 1980s, who wrote an almost adulatory account of him in *The Songlines*; and Lewis-Williams through the international best-selling author Graham Hancock, who recently lionised his breakthrough research into San art and San cultures in the book *Supernatural*.

Towering above all these larger-than-life figures was South African prime minister and philosopher–statesman Jan Christiaan Smuts, whose command of numerous fields of scientific knowledge is noted by Dubow. It was Smuts who recognised the veracity and importance of Dart's interpretations of the Taung skull; it was Smuts who patronised South African palaeontology and archaeology between the 1920s and 1940s, so giving it an international head start. It was Smuts who appointed Broom to a new position in the Transvaal Museum, thereby making possible the discovery of the first australopithecines to have been recovered since the 1924 discovery of the Taung skull, firstly in 1936 at Sterkfontein, and later of 'Mrs Ples', in 1947; and it was Smuts who financed Broom's 1946 monograph which finally convinced the world of the paradigmatic shift regarding human evolution first heralded by Dart.

If Smuts can be considered an intellectual giant of his era, he too was not free of the prevailing prejudices of his time that asserted the truth of white supremacy and black inferiority, which in turn justified white domination. While Smuts could outrage Europe by proclaiming Africa the birthplace of humankind, thereby rudely displacing it as one centrepiece of human evolution, he simultaneously sought to perfect a system of racial discrimination and racial subordination that went by the name of segregation, which systematically discriminated against and denigrated South Africa's blacks and out of which apartheid emerged. It is hugely paradoxical, from our current perspective, that Smuts could simultaneously entertain these two sets of ideas and beliefs. Part of this apparent confusion can be explained by reference to the growing 'poor white problem' in South Africa, which will be discussed a little later in this introduction. Part of it flowed from the perception which Dubow discusses in his chapter that a belief in a linked evolutionary past is very different from a universal sense of humankind. Still another part of it can be explained by related intellectual and biological theories of degenerationism. As is noted in the introductions to Parts 4 and 5, Smuts shared the view of many of his scientific contemporaries that humankind had not only evolved in the African interior, but had subsequently degenerated there as well. Having first rejected what later was called creationism, he now resolved the intellectual dilemma with which this presented him over the issue of race, by embracing the idea of degenerationism: while those early humans first stagnated and then degenerated inside Africa, the higher stages of evolution were accomplished by those other enterprising hominids who migrated out of Africa into Europe and Asia. All subsequent achievements and advances which occurred on the African continent were then credited to a flow of reverse migrations from the north as, for example, Smuts believed was the case with regard to the early civilisation of Mapungubwe, in the northern region of South Africa. For him (and for Dart), Mapungubwe was the work of superior immigrant 'Boskop Man' (who turned out to be a fiction) who was allegedly responsible for the fine line engravings found on rocks in various parts of South Africa, before this tradition degenerated in turn (Dubow 1995). This indeed was the fate of all initially superior black immigrant groups from the north, which presumably made them unworthy of serious attention or study.

The segregationist thrust in Smuts's thinking was driven forward and buoyed up by the threat to white domination supposedly presented by poor whites. This theme is discussed in the introduction to Part 5 and in Chapter 15 by Tim Clynick. Poor whites were perceived by Smuts and most of his political generation as the soft underbelly of white supremacy. Already consigned to minority status compared to blacks, white South Africa considered the loss of a significant number of its citizenry to poor-whitism (a condition into which a huge 50 per cent of Afrikaners were deemed to have sunk by the early 1930s) as a factor that would profoundly imperil white dominance. A host of projects were initiated to provide protected employment to poor whites, which almost always simultaneously discriminated against blacks. The flagship of these was the Hartbeespoort Dam, an irrigation scheme directly opposite the Cradle, which is discussed in Chapter 15. The job reservation and white labour policies at the centre of the Hartbeespoort and other schemes gave extensive artificial protection to whites in the face of black competition. Herein lies one of the motor forces of segregation. As noted in the introduction to Part 5, a man of Smuts's intelligence must have recognised the inconsistency of his intellectual and political positions on white racial superiority. The core conundrum which this raised was as follows. If whites were naturally so superior to blacks, why were they so in need of artificial racial protection? It seems possible that a hidden or even subconscious thread linked and allowed Smuts to reconcile these opposing positions. That link or thread was degenerationism. A core prejudice about poor whites was that they were experiencing a process of physical, mental and moral degeneration (although critics were more likely to blame inbreeding and isolation on the frontier than the hot African sun). It was this that placed them in need of protection and segregation. Poor whites, it was constantly averred, needed to be reclaimed or saved.

The notions of degenerationism which pervaded the thinking of Smuts and much of the scientific establishment of his time possessed not the slightest scientific foundation, but they both shaped and constricted the scholarly and scientific agenda up until the Second World War. As Esterhuysen remarks in her introduction to Part 3, despite the growing list of fossil discoveries in the Cradle and elsewhere in South Africa in the 1930s and 1940s, the technological revolution associated with stone tools was still believed to have occurred outside of South Africa and then to have been imported back into the region.

It was only in the 1940s and 1950s that the lie was given to these assumptions. As Kevin Kuykendall and Goran Štrkalj observe in Chapter 2, the study of palaeontology depended on certain social and scientific prerequisites to enable it to develop. One of these was the geological data and understanding to be able to secure a rough dating of fossils. As it transpired, fossil dating was wildly adrift up until 1959. Prior to that, as Kuykendall and Štrkalj point out, the whole of human evolution had been collapsed into a period of 1 million years. Then, with the arrival of potassium argon dating, that timescale was more than doubled.

From this point on, as is discussed in some detail in Chapters 2 and 3, the pace of fossil discovery accelerated as Broom, Robinson and Brain laboured at Sterkfontein and Swartkrans and Tobias and Hughes at Sterkfontein. Now, instead of degeneration, a picture of hominid efflorescence was revealed.

Other misconceptions have also been disproved. DNA shows the origin of modern humans to have been in Africa; and archaeological studies are pushing back the origins of modern human behaviour through the discovery of early signs of art, and signs of personal and other adornment which are taken to indicate the development of symbolic thought (see Chapters 6 and 7 by Amanda Esterhuysen and Lyn Wadley). Genetic science both reinforced some of the findings of physical anthropology and hominid palaeontology, and latterly totally revolutionised the field. In the 1940s and 1950s, as Himla Soodyall and Trefor Jenkins note in Chapter 4, Zoutendyk and Shapiro demonstrated that the San and Khoi were Africans and not influenced by Asians, as Broom and Dart had claimed. In the late 1980s Cann, Stoneking and Wilson published their path-breaking research into mitochondrial DNA, which claimed that all living peoples in the world could be traced to a common ancestor who lived in Africa about 200 000 years ago. This 'Out of Africa' theory was amplified by further studies, including those conducted by Soodyall and Jenkins at the University of the Witwatersrand, which revealed that some of the oldest DNA lineages found in the world's population are conspicuous among Khoisan populations. Y-chromosome studies confirmed and expanded on these conclusions. Other findings from DNA studies have since helped to suggest a recent and conservative date for the acquisition of speech by humans, and physical reasons (and dates) for the later growth of the human brain. In sum, a massive advance in our understanding of the evolution of humans has taken place.

The same notions of racial backwardness and even degeneration blighted research into and understanding of later periods of the Cradle's history. Up until the early 1970s, most South Africans, white and black, believed that Bantu-speaking peoples immigrated into South Africa some time between the thirteenth and fifteenth centuries (that is, around the same time as the first whites). Much as in palaeontological research, this belief was allowed to endure by the lack or unavailability of dating mechanisms for societies in the more recent past. In 1972/73, as the introduction to Part 4 and Chapter 9 by Tom Huffman show, the first dates for previously unidentifiable settlements were given. The most important of these, uncovered by Revil Mason, came from Broederstroom, which abuts the Cradle. At a stroke, the timescale of Bantu-speaking settlement of South Africa was pushed back an astounding one thousand years. South Africa's Early Iron Age had been discovered on the doorstep of the Cradle. At much the same time, the density and scale of Later Iron Age communities in South Africa came to be recognised, through aerial photographs and the archaeological digs of Revil Mason. These had been recorded by the first European travellers in the interior of South Africa, who spoke of impressive Tswana cities numbering 15–20 000 souls which rivalled Cape Town in size, but subsequently their observations had been suppressed or ignored. One major site in this complex was Olifantspoort in the Magaliesberg valley, whose rise and fall, together with that of other Tswana cities, are described by Simon Hall in Chapter 10. No other African cities of this size have been identified and investigated in any part of South Africa. Even now, though, they do not attract the attention they deserve.

An added damper on research into the recent history of South Africa's black majority was the doctrine of apartheid, which became official government policy in 1948 and blighted South Africa for forty years. Accompanying apartheid was the policy of Christian National Education, which denied evolution, reasserted special creation, affirmed Afrikaners to be God's chosen race in South Africa, and infantilised African society and culture. Important African achievements of the past were neither imagined nor sought. While major strides continued to be made in documenting evolutionary development in the Cradle and beyond by Tobias, Hughes, Brain and others, the Afrikaner establishment, the Afrikaner churches and most of the Afrikaner universities insisted on denying or ignoring them. A curious kind of national split consciousness was the result. We return to this theme in the Epilogue to this volume, which considers the relationship between politics and science in South Africa after 1945.

CHAPTER 1

WHITE SOUTH AFRICA AND THE SOUTH AFRICANISATION OF SCIENCE:
HUMANKIND OR KINDS OF HUMANS?

Saul Dubow[1]

The opening address to a conference on the 'African Renais-Science', held in Durban in 2002, was delivered by the world-renowned anatomist Phillip Tobias, who spoke on 'Africa: The Cradle of Humanity'. In a brief survey of hominid development and human evolution, Tobias lent his personal authority to what is rapidly becoming part of a key set of tropes in the discourse of the African Renaissance: that *Australopithecus africanus* is the progenitor of all living humans; that 'Africa gave the world its first culture'; and that hominids in southern Africa were 'not a southern African aberration, but a pan African revelation' (Tobias 2002: 6, 10).

It is not without irony that one of the central tenets of the new African Renaissance amounts to a neat inversion of what was once a core component of the country's tradition of scientific racism and white South Africanist assertion (Dubow 1995). In segregationist South Africa the discovery of australopithecines and other fossil remains was commonly held up as proof of underlying typological racial difference in modern human beings: it emphasised evolutionary diversity over unity, and it proclaimed scientists like Raymond Dart and Robert Broom as exemplars of colonial South African scientific achievement.

Until quite recently, insistence on a linked evolutionary past did not necessarily imply a belief in a common humanity: humankind, it was widely assumed, was composed of different kinds of humans. Paradoxically, Dart's revelation that the African continent harboured the origins of humankind is now actively seguing into a romantic vision of indigenous initiative and achievement. The 'back to Africa' model which portrays the African continent as the original cradle of civilisation serves, in the words of Dialo Diop, as a key prop of 'the cultural foundation of the African renaissance'. UNESCO's proclamation of the fossil-bearing sites at Sterkfontein as a World Heritage Site in 1999 has given an international seal of approval to a narrative that is also being marketed as the basis of 'Afrikatourism' (Diop 1999: 3–4, 8; see also Moaholi 1999).

Jan Smuts and the 'Great Divide'
It is well to remember that to proclaim a linked evolutionary past is quite different from asserting a universal sense of humankind. Indeed, for much, if not most of the past century, the emphasis has been on inscribing white South Africans into the story

of scientific progress and civilisation. Blacks, by contrast, have been largely ignored or else denigrated for their alleged primitiveness and/or the pre-scientific nature of their cultures.

In South Africa, scientific and other forms of knowledge have always been bound up in views of national identity and belonging. In the early nineteenth century, intellectual debates and the formation of scientific institutions were linked to the exercise of civil rights, the assertion of colonial respectability, and the promotion of a form of civic nationalism consonant with a liberal Cape colonial identity. During the first half of the twentieth century, scientific culture moved beyond its close association with the Cape to become one of the central supports of the ideology of broad South Africanism, associated in particular with the politician and public intellectual Jan Smuts, and his deputy, Jan Hofmeyr. While disdaining ethnic nationalism of the Afrikaner variety and stressing the virtues of internationalism, South Africanism took for granted the superior attributes of Western civilisation, rationalism and progress. Implicitly or explicitly, intellectual achievements were publicly celebrated in order to construct and defend an ethnically inclusive but racially exclusive white nation state.

Jan Smuts was undoubtedly the most cogent champion of South African science and its role in the creation of nationhood. His understanding of this link was bound up in his theory of holism which located the local and the particular as an aspect of the transcendent and the universal. Smuts's 'patriotism of place' was expressed through his love of nature, his expertise in botany, and the key role he played as a patron of ecological thinking. This newly developing field of interdisciplinary study, in which South Africa excelled and had much to contribute, helped to cement the country's position within the evolving network of imperial or, more properly, Commonwealth knowledge systems (see Dubow & Marks 2001; Anker 2001).

In 1925 Smuts served as president of the South African Association for the Advancement of Science, an honour that recalled the age when influential political figures were routinely invited to act as intellectual patrons of public bodies. His address to the Association, on the topic of 'South Africa in Science', considered 'the South African point of view' by sketching out the contribution the country could make to the sum total of human knowledge. Here Smuts declared his desire to reorient science away from the 'habits of thought and the viewpoints characteristic of its birthplace in the northern hemisphere', pointing instead to the country's distinctive position on the African subcontinent and in the southern hemisphere more generally. Employing a geological metaphor, Smuts posited Africa as the 'great "scientific divide" among the continents, where future prospectors of science may yet find the most precious and richest veins of knowledge' (Smuts 1925: 3–4).

Fig. 1.1 Dart with 'Taung child'

Smuts began by describing the implications of the 'Wegener hypothesis', named after the German geophysicist whose *Origin of Continents and Oceans* (1915) pioneered the theory of continental drift. His innovative use of Wegener's ideas, which were far from commonly accepted at the time, served as a fitting leitmotif. It allowed Smuts to posit Africa as the southern hemisphere's 'mother continent' from which South America, Madagascar, India and Australasia had originally split away or 'calved off'. By placing southern Africa at the centre of this 'great "divide"', it now became possible to correlate scientific developments across a range of disciplines in new and creative ways, with potentially far-reaching implications for 'universal science' (Smuts 1925: 4–5).

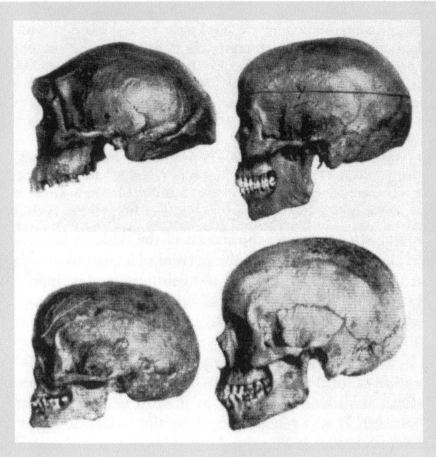

Fig. 1.2 Four ancestral types R. Dart

In the field of geology, for example, several Cape formations appeared to be mirrored by equivalents in India and South America; the pattern of mineral deposits, such as diamonds and coal, illustrated a similar symmetry (Smuts 1925: 5–6). Botany and palaeo-botany posed especially interesting comparative questions. Whereas most South African flora was evidently of tropical origin, the south-western Cape was characterised by a distinct temperate flora which could not – as was then widely believed – have derived from northern Europe. Instead, Smuts suggested, Cape flora might have come from ancient Gondwanaland, the continent which was now largely covered by the South Atlantic Ocean but which had once encompassed much of Africa, Australia, India, South America and Madagascar. This theory could well explain botanical affinities between flora in South Africa and in other countries of the southern hemisphere (Smuts 1925: 6–7).

Smuts proceeded to apply the Wegener hypothesis to zoology, climatology and meteorology, drawing on the evidence of past ice ages to explain rainfall patterns and desertification. He also extended his field of vision to the heavens, referring to the pioneering work of Lacaille, Henderson and Gill. South African astronomy, Smuts reminded his audience, 'has the distinction of being responsible for the determination of both great astronomical standards of measurement – the distance of the sun and the distance of the fixed stars'. Finally, Smuts paid tribute to the new field of human palaeontology which, he said, was bringing South Africa 'right into the centre of the picture' (Smuts 1925: 14–15, 16).

In the two preceding decades palaeontology had been given strong impetus, especially following AC Haddon's 1905 address to the joint meeting of the British and South African Associations for the Advancement of Science. The pioneering Cambridge anthropologist, who had risen to international prominence as the organiser of the 1898–99 Torres Strait expedition, used the occasion to call for more work on the 'interesting subject of comparative physiology'. Duly inspired, the director of the South African Museum, Louis Péringuey, focused attention on its extensive collection of indigenous skulls, skeletal material and stone artefacts. He also inaugurated a major project to model life-size casts from surviving 'Bushmen' and 'Hottentots'. The discovery in 1913 of a large-brained and thick-boned fossil skull, which Robert Broom described as 'Boskop Man' (*Homo capensis*), opened a new era in palaeontology: most notably, it lent credence to the suspicions of those nineteenth-century investigators who had long suspected that human origins in South Africa might have a deep past.

Broom hypothesised that 'Boskop Man' (a highly racialised formulation and one that has long since been discredited) might be a direct ancestor of the 'more or less degenerate Bushman of recent times' and he posited connections with Neandertal and Cro-Magnon skulls found in Europe. Further discoveries, such as that of 'Rhodesian

Man' in 1921, sharpened public interest in the prehistoric hominids of southern Africa and fed into a growing fascination with human origins and race typology. This momentum was sustained after the First World War by the appointment of professional comparative anatomists at the university medical schools of Cape Town (Matthew Drennan) and Johannesburg (Raymond Dart). Having been based in the museum sector up to this point, physical anthropology now had a firm institutional foothold in the universities (Dubow 1995: 39).

This presence was powerfully entrenched after 1925 when Raymond Dart announced in the journal *Nature* that a newly discovered fossil skull, which he dubbed *Australopithecus africanus* ('the Man-Ape of South Africa'), marked a crucial transitional 'missing link' in hominid evolution. Dart suggested that *Australopithecus* might vindicate Darwin's prediction that 'Africa would prove to be the cradle of mankind' (Dart 1925: 195–199). His audacious claim was greeted with a mixture of scepticism and outright hostility on the part of leading international experts who were disinclined to accept that either South Africa, or its unknown tyro anthropologist, merited the honour that attached to such a major discovery. Two decades would pass before further discoveries, notably those made by Robert Broom at Sterkfontein, shifted the international consensus in Dart's favour.

In his 1925 address Smuts went out on a metaphorical limb by giving support to Dart's theory. He speculated that 'South Africa may yet figure as the cradle of mankind, or shall I rather say, one of the cradles?' (Smuts 1925: 17). In referring to 'cradles' in the plural form, Smuts probably had in mind the strength of rival claims, but he might equally have been gesturing towards the polygenetic or multilinear theories of evolution that continued to exert influence. He certainly lent his personal intellectual authority to the racialised form of comparative anatomy that Dart and his followers did so much to promote.

For instance, in arguing for South Africa's importance as a field of anthropological research, Smuts ventured the opinion that '[o]ur Bushmen are nothing but living fossils whose "contemporaries" disappeared from Europe many thousands of years ago'. As such, he said, they were analogous to the country's cycads which were fossil survivors of the botanical world (Smuts 1925: 16–17). The notions that the aboriginal races of the country represented the end of an evolutionary line, that Bushmen could be likened to the indigenous flora and fauna of the country, and that they should be preserved primarily as evolutionary curiosities, were certainly commensurate with racial attitudes during the segregationist era (see for example Gordon 1992 and Dubow 1996). This, too, formed part of South Africa's 'great divide'.

Knowledge and national identity

Broadly speaking, the relationship of knowledge to power, and its role in supporting claims to national identity, went through four identifiable phases from the early nineteenth to the mid-twentieth centuries. The first of these covers the period from the 1820s to the early 1850s. In this era, knowledge and knowledge-based institutions formed part of a drive to establish a middle-class civic order at the Cape that resisted autocratic colonial rule and asserted the rights owing to respectable citizens in a British colony of settlement. The second phase, which took in the remainder of the century, saw a growing convergence between knowledge-centred institutions and the constitutional prerogatives of a colony enjoying the fruits of self-government. The tensions between colonial nationalism and imperial assertion shaped this relationship until the outbreak of the South African War in 1899.

A third phase was entered into during the period of post-war reconstruction in the first decade of the twentieth century. The politics of knowledge now became closely associated with the spirit of reconciliation between Boer and Briton and were deliberately utilised in support of the ideology of South Africanism. Scientific and cultural bodies helped to give shape to the emergent white nation state by building up its intellectual infrastructure. The period from the outbreak of the First World War to the close of the second inaugurated a fourth phase in the politics of knowledge. Science was now used to underpin a sense of South African patriotic achievement within the broader context of Commonwealth belonging. The period also saw growing state interest in, and direction of, practical knowledge in the social and natural sciences.

Notwithstanding a number of pioneering enquiries into African societies in the early nineteenth century (typically conducted in the 'manners and customs mode' developed by earlier travellers, explorers and missionaries), there was not yet any concerted or systematic interest in the origins of humankind or the timescale of human habitation in Africa. This began to change at mid-century when a significant measure of self-government was ceded to the citizenry of the Cape. The public institutions of the Cape expanded during this period, and the emergent political culture encouraged formalisation of working agreements that had evolved between Dutch- and English-speakers over two generations.

The development of an anglophone colonial intelligentsia from the mid-1850s was immeasurably stimulated by the creation of the *Cape Monthly Magazine* (Dubow 2004). Modelled on British periodical literature, the *Monthly* served as a vital medium for the interchange of ideas. Its constant watchwords, of progress and improvement, were directed to the formation and expansion of literary and scientific institutions. As well as giving vent to new ideas generated in the metropole, the *Monthly* provided an opportunity for southern African-based contributors to report on local discoveries and to generate a continuous and multilayered conversational thread.

Inspection of the debates and discussions in the *Cape Monthly* from the early 1870s yields substantial evidence of the development of a visible – and invisible – colonial college of ideas. Its columns reveal how newly developing associations between geography, geology, palaeontology, and anthropology brought together fragmentary knowledge about landscape, topography, rock formation, stone tools and Stone Age cultures. Concerns with antiquity (geological time and human habitation) merged with studies of human and material culture (language, stone implements and rock art). Evolutionist ideas underpinned many of these associated discoveries. A growing awareness of, and concern with, racial difference gave these abstract ideas striking topicality and resonance.

For leading metropolitan scientists (like Roderick Murchison, Henry de la Beche and Richard Owen), geographical, geological and palaeontological knowledge together provided the means to assert symbolic order over the British dominions, as well as to recover the 'lost worlds' of the distant and remote prehistoric past and to revisit 'the ancestral stages of their own culture' (Stafford 1999: 317). For investigators based on the South African sub-continent, interest in geology was often conducted in the context of travel, exploration or prospecting. It offered a means of surveying the landscape and its inhabitants, of locating oneself in time and in place, and of laying claim to the country and its resources. As the importance of local discoveries received public recognition, so the distinctiveness of South African natural history became apparent. In due course individual achievements helped to foster local colonial identity and pride.

The idea that human habitation in South Africa might be measured in geological time engendered a spirited debate amongst Cape-based intellectuals.[2] The educationist and leading public figure, Langham Dale, who found and recognised a number of stone tools on the Cape Flats in the 1860s, introduced to readers of the *Monthly* the subject of archaeology and pointed to its importance in the understanding of the remote human past. He called it 'a new science, interposing between Geology and History' (Delta [Langham Dale] 1870).[3] One of the striking features of this new science as it developed on the African continent was the ambivalence many of its practitioners displayed when it came to pushing back the temporal frontiers of human habitation or ascribing creative agency to its indigenous inhabitants.

Dale was one of the first collectors to link the stone tools and pottery found at the Cape with living Khoisan peoples, yet he was reluctant to draw any firm conclusions: 'The stone-tools found in South Africa,' he wrote in 1870, 'do not contribute any evidence towards the solution of the state of primitive man; at the same time, they do not tend to discredit the notion of a remotely distant existence of mankind on the European continent.' As yet, there was insufficient evidence with which to construct a theory concerning the implements and pottery so far found in South Africa, and nothing to prove that they were 'prehistoric' (Delta 1870: 236–237; see also Goodwin 1928: 419–426).[4]

The difficulty in accepting a deep-time history for the peopling of the sub-continent centred on the implications this had for scriptural accounts of creation, as well as for ideas about the nature of primitive humans and their potential for civilised development. The absence of incontrovertible empirical knowledge in the 1870s (no evidence of fossilised human remains had yet been discovered) and the fact that competing theories of evolution had not yet freed themselves from the ooze of theological received wisdom, meant that the question of humankind's origins remained a matter for speculation. From the 1880s the idea of natural selection was increasingly deployed to link theories of evolution and progress to ideas of racial origin and notions of innate superiority. But although most of the elements were already available to effect this noxious synthesis – and were already evident in embryonic form – they had not yet coalesced into a clear ideological and explanatory framework.

Fig. 1.3 Smuts (centre) in the field with van Riet Lowe (left)

As well as being tied to problems of imperial governance and labour supply, the growth of expert knowledge about the land and its peoples was closely bound up with processes of colonial self-discovery and understanding. The urge to know about others was born of intellectual curiosity and the urge to constitute a sense of collective self. It also had a more instrumental dimension, namely, the power to identify, pronounce upon, and control South Africa's indigenous inhabitants. Many contributors to the *Cape Monthly* were pioneers in what would later become known as 'African Studies'. Amateur colonial experts took care to formulate and format their ideas so as to fit in with, and inform, universal Western schemes of knowledge. Theirs was the knowledge *about* Africans rather than the knowledge *of* Africans themselves.

Colonial nationalism and indigenous belonging

At another level, the colonial intellectuals who helped to define, delimit and describe the indigenous peoples of the country were becoming African. Their expertise was closely bound up with claims to be rights-bearing citizens of a country that they were making their own. To make this assertion entailed displacing, limiting or bypassing the prior ownership rights of black and brown peoples, while simultaneously demanding the rights of colonial citizenship within a wider British world. This is precisely what colonial nationalism involved.

Assertions of colonial nationalism could be more or less exclusionary. Within the rubric of its more liberal emanation – and subject always to acceptance of the principles of civilisation and the sovereignty of the Crown – British settlers, Afrikaner colonists and (in certain cases) blacks, too, could be accommodated. Colonial nationalist awareness was sharpened considerably over the course of the last quarter of the nineteenth century. The mineral revolution in the South African hinterland, the final conquest of African peoples in the sub-continent, and the failed efforts by the metropole to confederate the loose assemblage of republics and colonies in a South African dominion, all contributed to this process.

The South African War of 1899–1902 marked a damaging defeat for the delicate inter-ethnic alliance of moderate Afrikaner nationalists, anglophone liberals and enfranchised Africans. But the ideology of South Africanism emerged to inhabit the space left by a retreating imperialism and a temporarily broken republicanism. Geared to the needs of a unified white nation state, it stressed virtues of moderation and conciliation. Emphasis was laid on the need to create broadly based institutions and to nurture a shared national culture as a necessary counterpart to successful state-building. The constitutional politics of the Cape and the traditions of Anglo–Dutch partnership were revived and reconfigured to take account of new realities and a much expanded national entity.

A key South Africanist assumption was the notion that national sentiment could be accommodated within a wider sense of imperial belonging. Racialism, understood as antagonism between Dutch and British, was invariably deplored. The language of social Darwinism, which had previously insisted on the incorrigible degeneracy of Boers, now emphasised the common 'Teutonic' roots of English and Afrikaans speakers. Conversely, insistence on the implacable difference between whites and blacks helped to rationalise the need for systematic racial segregation.

Scientific and technical agencies, as well as professional bodies, made a significant contribution to South Africanist ideology and helped to embody its meaning. The notion of science as transcendent truth rendered it possible to cast the language of progress and universality within the imperial 'chain of civilisation'. One of the first national public institutions to be created after the South African War was the South African Association for the Advancement of Science. Just as its Australasian equivalent had been seen as a harbinger of the Australian Commonwealth, so the South African Association came to be seen as a step towards 'closer union'. The joint meeting of the South African Association with the British Association in 1905 exemplified the benign spirit of reconciliation in a constructive and benevolent manner. In fields like agriculture and veterinary medicine, the demonstrable benefits of scientific cooperation were eagerly taken up by ambitious Afrikaner modernisers and supported by politicians like Louis Botha, the first prime minister of the Union of South Africa, who personally exemplified the spirit of reconciliation.

Several leading mining magnates and politicians, who were admired and reviled in equal measure for their support of the imperialist cause, gave direct support to the ethos of South Africanism by promoting significant new cultural and scientific ventures. These ranged from museums and art galleries to the creation of the national botanical gardens at Kirstenbosch and moves to establish a national teaching university. Identification with the landscape was practically encouraged through the promotion of agricultural improvement, conservation and irrigation. Railways enclosed the outer reaches of the country within the realms of civilisation and became a preferred means of stimulating tourism and national awareness. Preservationist and conservationist movements led to the creation of the Kruger National Park and the consecration of Table Mountain as a symbol of national unity. Enthusiastic literary and photographic evocations of 'the veld' encouraged fresh appreciation of a landscape that had once been routinely dismissed as barren and featureless. These initiatives and responses, in which scientific awareness and aesthetic sensibility were freely interwoven, were all part of a concerted effort to encourage a sense of inclusive South Africanness.

If the South African War opened the way for South Africanism to take root as a political ideology, the First World War put it to the test. Participation on the side of the empire reopened wounds and provided Afrikaner nationalism and republican revanchism with fresh impetus. Hertzog was a clear beneficiary, but Smuts was able to consolidate and refashion his political base. Alternative visions of South Africanism were offered by the two leaders: Hertzog stressed national determination (though stopping short of republicanism) while Smuts became increasingly dependent on the imperial connection. The slow haemorrhage of support to the Nationalists was partially compensated for by Smuts's growing international status and the transmutation of empire into the idea of Commonwealth. Over the next quarter of a century, South Africanism of the Smutsian variety would remain an enduring political ideology.

Inter-war state-sponsored programmes in the social sciences and humanities suffered from underfunding, lack of effective intergovernmental coordination and an absence of clear objectives. Yet the ethos of South Africanism was fully reflected in efforts to disburse funds on a non-partisan basis (that is, to English as well as Afrikaner institutions) and in a growing emphasis on research with demonstrable South African relevance. Palaeontology and prehistory were natural beneficiaries of such interest. In a major summary of developments in the field, the pre-eminent Cambridge prehistorian Miles Burkitt observed that 'the study of prehistory in South Africa [is] so interesting and important to members of the various nationalities who have chosen to make that beautiful country their home, and who are in the process of forming a new South African nation within the British Empire' (Burkitt 1928: 1).

Developing international scientific contacts and networks, especially in the fields of comparative education and psychology, encouraged the promotion of South Africa as an international 'laboratory' for comparative investigations into race and culture. The eventual withdrawal of Carnegie funding to government departments led to the creation of a National Research Council and Board in 1938. This was launched amidst a fanfare of patriotism and hailed by JH Hofmeyr as a South African 'parliament of research'. Yet the new body did not deliver on its promise to increase research funding, and many of its operations were interrupted by the Second World War.

The limited success of state-managed intellectual activity was compensated for by research conducted in the university sector, as well as in semi-autonomous institutes like Onderstepoort Veterinary Institute (established in 1908), the South African Institute for Medical Research (1912), and the Carnegie-funded Bernard Price Institute for Geophysical Research (1937). In 1929, on the occasion of the second visit to South Africa of the British Association, Hofmeyr and Smuts trumpeted the 'South Africanisation of science'. A great deal was made of the expansion in tertiary education and the growth of research infrastructure in the period since the British Association's previous visit in 1905. Discoveries of national – and international – significance in the field of prehistory and physical anthropology generated a great deal of publicity.

Smuts was especially adept in placing South African scientific achievements in a broader perspective and harnessing them to the purposes of statecraft. His authority as an international statesman was enhanced by the publication in 1929 of his personal philosophy of holism which sought to synthesise scientific knowledge and human potential in terms of organic unity and evolutionary progress. It also provided a ready analogy for his views about political interdependence: in the Smutsian universe, small nation states like South Africa could expand rather than dilute their identity by association with the Commonwealth, the League of Nations, or the United Nations – with all of which bodies Smuts had a close involvement at key moments in their development.

The 1929 meeting of the South African and British Associations revealed a less deferential attitude to the empire on the part of local scholars, as well as exposing underlying ethnic tensions. From a religious and ideological perspective, and also for reasons of practical politics, Afrikaner nationalists were angered at the way in which scientific universalism was equated with membership of the Commonwealth. Strong exception was taken to EG Malherbe's presentation of poor-whitism at this meeting in terms of Afrikaners' psychological feelings of inferiority. The Afrikaner nationalist presence in science was still undeveloped, but there were plenty of intimations that this was a field in which the *volk* was beginning to think about asserting itself. It would take another political generation for the anglophone dominion of knowledge to feel this challenge in earnest.

Colonial knowledge producers presumed the universality of Western scientific knowledge and sought to root its ideas, institutions and systems in an African context. Unless local knowledge and belief systems could be translated and absorbed – with or without attribution, and always selectively – within a Western framework, these tended to be disparaged as quaint (at best) and irrelevant or irrational (at worst). The power to declare Boer or African folk knowledge as valid or invalid was assumed to be the sole prerogative of Western science.

This should not come as a surprise, but it was not solely a function of racism or prejudice. Colonial patriotism and pride depended on the capacity to demonstrate that the local and the specific formed part of a larger, universal, scientific scheme. There were, moreover, close links between developments in colonial naturalism and white colonial *nationalism*. To know the land and to conceptualise its peoples was to assert cognitive power and to proclaim a custodial or proprietorial sense of ownership. It was, in a sense, an assertion of acquired indigeneity.

'Little Foot' from Sterkfontein (Stw 573)

FOSSILS AND GENES:
A NEW ANTHROPOLOGY OF EVOLUTION

Trefor Jenkins

The 'Cradle of Humankind' is so named because of the stunning palaeoanthropological discoveries which have been made in that area, firstly by Robert Broom in 1936, and over the subsequent seventy years by a number of other scientists and their dedicated helpers. The hominid fossils found there cover a time span of from 3 or 4 million years ago to perhaps 500 000 years ago and are well described by Kevin Kuykendall in Chapter 3 of this section, as are the plants which populated this area over the same period of time and which are discussed by Marion Bamford in her presentation of the available evidence in Chapter 5. Little attention is given to the faunal (that is, non-hominid) remains, but these have been studied by Basil Cooke, Lawrence Wells and Bob Brain over many years, and by younger researchers as well.

In Chapter 2, Kevin Kuykendall and Goran Štrkalj introduce the reader to the birth of palaeoanthropological research in South Africa and its development as a scientific discipline in a country 'that was struggling for political recognition by the West, and during which time the foundations of the conservative apartheid political system, for which South Africa became infamous, were formulated.' Dart's discovery, description and interpretation of the 'Taung child' in 1925 (Taung is situated in the North West province, some 500 kilometres west of the Cradle) flew in the face of the orthodox view of human origins held at that time, and it took another twenty years before Dart was fully vindicated. In 1936 Broom discovered an adult fossil skull (*Australopithecus transvaalensis*) at Sterkfontein; it was initially named *Plesianthropus transvaalensis* and was considered to be an adult version of Dart's australopithecine child from Taung, thereby contributing to the vindication of Dart's initial 1925 claim. Further discoveries by Broom and his associates at Kromdraai and Swartkrans, situated close to Sterkfontein in what is now known as the 'Cradle of Humankind', provided additional insights into the anatomy of humankind's early ancestors.

Broom, discussing his discoveries at Sterkfontein, Kromdraai and Swartkrans, wrote in one of his last books, *Finding the Missing Link* (1950), that more discoveries would be made and that 'our caves when fully worked, as they will probably be in the next fifty years, will almost surely give us many more genera' (Broom 1950: 57). Fifty-six years later, the sites opened up by Broom have still not been fully excavated, and other sites in the Cradle are being explored; exciting discoveries continue to be made.

For the hundred years after Darwin enunciated his theory of evolution by natural selection, and suggested that humans had originated in Africa, the primary evidence for human evolution came from the study of fossils. Because many of the first dis-covered and most informative hominid fossils came from the 'Cradle of Humankind' and the interpreters of them lived and worked in South Africa (Dart, Broom, Robinson, Schepers, Brain and Tobias), the foundations were laid for the blossoming of palaeo-anthropology in this country. The University of the Witwatersrand, which produced dozens of PhD graduates in the field, together with the Transvaal Museum (which now forms part of the Northern Flagship Institution), attracted scientists from all over the world who wished to study the hominid remains, their environment and their cultural artefacts.

The first two chapters of Part 2 have been written by former students of Phillip Tobias who, within a few years of his appointment as Raymond Dart's successor as head of the Department of Anatomy at the Wits Medical School, began a long-term excava-tion of the Sterkfontein cave; for the first twenty-seven years or so (from 1966 to 1993) the fieldwork was supervised by the late Alun Hughes and, when he died in 1993, his place was taken by RJ Clarke.

It was Bob Brain (1958) who gave the first detailed description of the geological structure of the Cradle's cave system before beginning, in 1965, his intensive study of the Swartkrans site, which extended over the ensuing twenty-one years. It was 'very much a family enterprise, in which I have had the continual support of my wife, Laura, and of each of our children, on tasks varying from the excavation of fossils, their preparation, cataloguing and identification, to experiments with bones in camp-fires' (Brain 2004: 2). Included among his many finds was the first demonstration of the use of fire by hominids about 1 million years ago; it is probable that those fires were gathered from natural lightning-induced blazes until *Homo erectus* devised ways of making fire at will, perhaps 500 000 years before the present (Brain & Sillen 1988).

Readers with a literary bent may have encountered Bruce Chatwin's account of the discovery of the first evidence of burnt bone at Swartkrans on 2 February 1984 (Chatwin 1987; Shakespeare 1999: 2–3). After lunch on that day, Chatwin, with Brain and the site foreman, George Moenda, had been digging 'close to the west wall' when Moenda found a fragment of antelope bone which was 'speckled with dark patches, as if burned'. Subsequent excavations showed that fires had been repeatedly made on the dif-ferent floors over a period of many thousands of years and that fire management was mastered between layers dated to about 1 million years and 500 000 years ago, by which time the hominids may have eventually developed ways of making fire at will; the early efforts were those of *Paranthropus robustus* while the later ones were those of *H. erectus* (Brain 2004; Grine 2004).

Chapter 3 by Kevin Kuykendall does not attempt to cover in detail all the hominid fossil finds in all of the cave sites in the 'Cradle of Humankind'. His focus is on the discoveries made at Sterkfontein, Kromdraai and Swartkrans, and on the ways in which evolving research techniques are enabling palaeoanthropologists to provide increasingly precise dates and descriptions of their finds. His fine discussion of the popular concept of 'missing links' demonstrates clearly that the concept is not only unhelpful but, in fact, erroneous.

The question of how modern humans evolved is far from settled, despite the many fossil finds and extensive research work described by Kuykendall. In a recently published volume, *The Complete World of Human Evolution*, Stringer and Andrews (2005), two eminent British palaeoanthropologists, review the events surrounding the discovery of the first *Australopithecus africanus* specimen by Raymond Dart in 1925, and the problems Dart experienced in gaining acceptance of his views of the skeleton's place in human evolution. Although Dart appeared to be vindicated ten to twenty years and more after his initial announcement of the find, Stringer and Andrews (2005: 125) claim that '[t]he wheel has turned almost full circle and many experts now echo the doubts raised in 1925 that these enigmatic creatures may only represent an extinct side-branch in the human evolutionary story'. They speculate that *Homo sapiens* may, rather, be descended from *Australopithecus afarensis*, who lived 3 to 4 million years before the present, possibly via *Homo habilis, Homo ergaster* and *Homo heidelbergensis*. In fairness to Dart, however, the chart which is displayed in this book shows that such an evolutionary path is dotted with many 'question marks'!

The naming of the species *Homo habilis* in 1964 by Leakey, Tobias and Napier (1964) marked the beginning of a new era in which the African continent was shown to be the place of origin of the earliest humans (Stringer 2006). During the last twenty years numerous fossils and artefacts have been found which have confirmed the African origin of humans. The techniques of palaeoanthropology have now been augmented by the research methods available to scientists working in the field of molecular genetics to produce increasingly detailed and accurate descriptions of the evolutionary development of humankind. With the advent of molecular genetics in the late 1970s it was possible to envisage that population or evolutionary genetic studies of living peoples would be able to contribute to the study of the origins of humans, as well as to the reconstruction of their migrations. In a ground-breaking article Cann, Stoneking and Wilson (1987), using the global distribution of mitochondrial DNA (mtDNA) variation, presented a genetically-based 'Out of Africa' theory of human evolution; the use of Y chromosome DNA, autosomal DNA and X chromosome DNA variation has provided convincing support for this theory. This exciting work, a valuable addition to the painstaking efforts of palaeoanthropologists over nearly a century, has given a new dimension to the study of human origins and evolution (Soodyall 2006). I hesitate to quote the somewhat irreverent opinion of the geneticist and gifted populariser of science, Steve Jones, who wrote: 'The fossil record will never give us the complete history of human evolution, but it can give us dates and places which genes can only hint at. It is worth glancing at the bones before staring at the molecules' (Jones 1994: 123). Molecular anthropologists have, in recent years, expanded their studies of genetic variation among many populations of the world; they are motivated by the knowledge that whereas a fossil human may, or may not, have left any descendants, living peoples must have had ancestors. With this reassurance, researchers around the world are studying the genetic variation among representative populations and are employing increasingly sophisticated statistical methods for analysing the data and reconstructing the evolutionary history of our species. Using the fossil record, it was estimated that humans and chimps had shared a common ancestor about 15 million years before the present; using molecular studies on proteins, the date dropped to 5 to 7 million years, a range amply confirmed by subsequent molecular genetic studies.

What are the prospects of obtaining DNA for analysis from fossilised specimens? This has been achieved, but by only very few laboratories. Mitochondrial DNA has been the DNA of choice because there are many hundreds of copies of the mtDNA molecule in every nucleated cell in the body, including bone cells. In every cell there are, in comparison, only two copies of the nuclear genome. Mitochondrial DNA sequence data have been obtained from four Neandertals and the findings fall outside the range of modern human mtDNA variation, showing that Neandertals diverged from the line leading to modern humans some 600 000 years ago (Krings et al. 2000). Soodyall and Jenkins (Chapter 4) review this work and, in addition, draw attention to the ways in which molecular genetics will shed light on the evolution of specific human traits, like speech, which would not be amenable to studies of the fossils.

CHAPTER 2

A HISTORY OF SOUTH AFRICAN PALAEOANTHROPOLOGY

Kevin Kuykendall and Goran Štrkalj

In February 1925 Raymond Dart, a 32-year-old professor of anatomy from the then obscure provincial University of the Witwatersrand, Johannesburg, announced the discovery of a new species of fossil ape-man. This discovery would prove to be one of the most important in the history of palaeoanthropology[1], the scientific discipline which studies human evolution. It would also establish South Africa as the place where some of the major phases of human evolution had taken place, and the University of the Witwatersrand as one of the most prestigious centres for research on human origins. However, all this would only happen much later, mainly because Dart's interpretation of the fossil was generally accepted only a quarter of a century after his initial announcement. To understand why this was the case, one must look at the early history of palaeoanthropology, its state at the beginning of the twentieth century, and the social context within which this science developed and was accepted or rejected. Even today, in a number of societies the concept of evolution, and especially human evolution, is taboo.

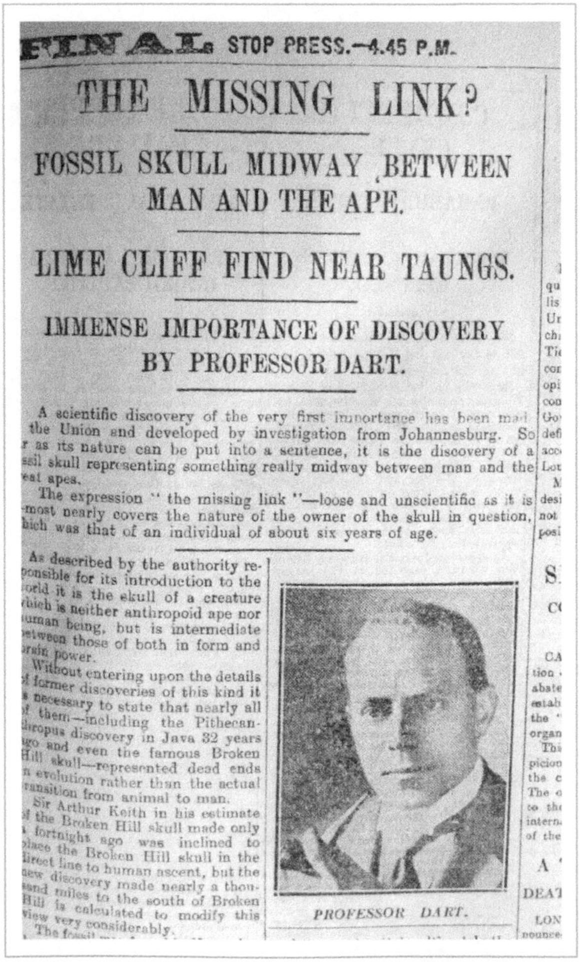

Palaeoanthropology as we know it today is a complex discipline which could not have emerged without certain

THE MISSING LINK?

FOSSIL SKULL MIDWAY BETWEEN MAN AND THE APE.

LIME CLIFF FIND NEAR TAUNGS.

IMMENSE IMPORTANCE OF DISCOVERY BY PROFESSOR DART.

Fig. 2.1 Original news report in the Johannesburg Star *of Dart's announcement of the 'Taung child' (February 1925)*

social and scientific prerequisites. Any science can develop only in a society where scientific research is encouraged and free thinking is tolerated. The freedom to challenge old views seems particularly important for palaeoanthropology, as the claim that humans had not been created by a divine power but had evolved, through the work of natural processes, from an animal ancestor, was potentially more subversive than the claim that the planet Earth was not at the centre of the universe.

This does not mean, however, that science necessarily flourishes only in democratic societies. From the autocratic rule of Ptolemy I in Hellenistic Egypt (when scientific authorities from all over the world converged on Alexandria) to the totalitarian states of the twentieth century, scientific achievements were seen as a means of promoting the state and its interests. This is probably the main reason why palaeoanthropology thrived from the 1920s onwards in South Africa, a society where up until recently evolutionary ideas were looked upon with suspicion. Paradoxically, while during the period when the National Party ruled the country evolution was not taught in schools, South African scientists were involved in some of the most important breakthroughs in the field of human evolution studies. Students of human evolution, therefore, seem to have been tolerated and sometimes even supported by the state as their accomplishment might contribute to the country's prestige and international recognition.

The rise of palaeoanthropology as a science

Palaeoanthropology is essentially a multidisciplinary endeavour, and could not arise until related sciences (above all anatomy, biology, geology and archaeology) had reached a certain level of development. While modern anatomical science emerged in the sixteenth century through the work of Andreas Vesalius and his followers, other scientific disciplines emerged only later. The biblical account that the world, humanity and all other species had been created by God some 6 000 years ago was generally accepted in Western society as the literal truth about world history until the nineteenth century. The idea that humans had evolved could only be considered by the broader scientific community once this biblical world view had been abandoned; only then did the search for ancestral human fossils begin.

The preconditions for the rise of modern palaeoanthropology seem to have been first fulfilled in the nineteenth century. The year 1859 might be viewed as the symbolic beginning of modern palaeoanthropology, as three important events took place then. Firstly, Charles Darwin published his masterpiece *On the Origin of Species*. Although evolution had been discussed previously and Darwinism itself would be questioned later on, the publication of this seminal work meant that the idea of evolution was here to stay. One of the reasons for the success of Darwin and his followers lies in the fact that they managed to gather a vast body of evidence supporting the view that evolutionary change actually happened. Darwinism subsequently became the ruling paradigm in the biological sciences.

Secondly, the strange claims of the French customs official Jacques Boucher de Perthes were vindicated by two English scientific authorities, Evans and Prichard. Boucher de Perthes maintained that the broken stones he had found in the Somme Valley were not produced by natural forces, as previously believed, but represented tools of early humans who had lived at the same time as the prehistoric animals found in the same geological deposits. At that time many scientists still adhered to the dictum of Georges Cuvier, a superb anatomist, palaeontologist and anti-evolutionist who had died half a century earlier, that human fossils did not exist. Thanks to Boucher de Perthes the idea of human antiquity became accepted or at least seriously considered by many, in spite of the fact that no human fossils were recognised at the time.

Fig. 2.2 Boucher de Perthes

Thirdly, Paul Broca, the famous French neuroanatomist and surgeon, established the first anthropological society in Paris. It was approved by the government with the proviso that a police official should sit in on all the meetings – anthropology[2] and evolution were apparently still seen as subversive subjects that questioned the old values regarding the traditional place of humans in nature, and could pose a danger to the stability of the state. Anthropological societies and university departments providing institutional support for the study of human evolution would soon appear in many other European countries.

Early work on human evolution focused on comparative anatomy and embryology. Similarities between humans and apes led many scientists to the conclusion that humans had evolved from an ape-like ancestor. The main aim of the research was to demonstrate that evolution had occurred, rather than to show how it had happened. The latter was difficult to achieve because of the paucity of available fossils. In this early period no one was systematically searching for hominid[3] fossils anywhere in the world, and the earliest finds were random discoveries by miners and other non-professionals.

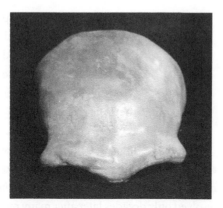

Fig. 2.3 Neandertal fossil cranium from Feldhofer Cave, Germany

Interpretations of such fossils varied widely because there was no established scientific context for their analysis, but the prevailing view was that the large human brain – the seat of our intellectual and philosophical abilities – evolved first, and other features such as walking upright on two legs (bipedalism) followed later.

The first hominid fossils found and recognised as such were those of the Neandertals in Germany; they were found three years before the publication of *On the Origin of Species* (two earlier finds were not recognised as Neandertals until the late 1880s). Soon, more Neandertal finds were made in other parts of Europe such as Belgium, France and Croatia. The Neandertal fossils induced heated discussions in scientific circles on various issues, including their authenticity as fossil humans. Eventually the majority of scientists accepted that they represented an extinct group of human beings, although they were viewed as a sideline branch rather than the evolutionary 'missing link' between apes and humans. In 1907 a robust lower jaw with teeth very much like those of modern humans was unearthed in a sandpit near Heidelberg in Germany. The jaw was older than any of the Neandertal remains. It was interpreted by the anatomist Otto Schoetensack as a possible precursor of Neandertals and christened *Homo heidelbergensis*. We now know that these fossils occupy a very late position in human phylogeny[4] – though they were considered to be ancient at the time of their discovery.

The first scientist to search intentionally for fossil human ancestors – the so-called 'missing links' – was the Dutch anatomist Eugene Dubois. Frustrated by the fact that no one would finance his expedition to the tropics, where he hoped to find fossilised remains of 'ape-men', Dubois joined the Dutch army and left for the Dutch East Indies (later Indonesia). After a few years of futile searching, he found a thigh bone in 1890 which was very much like that of modern humans. A year later, at the same site, he

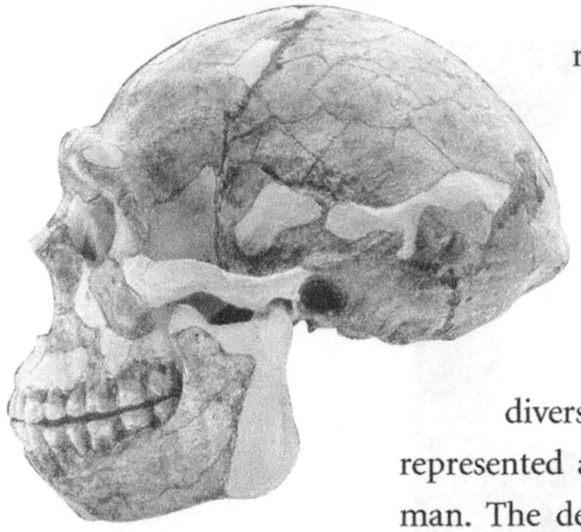

Fig. 2.4 Pithecanthropus *cranium*

recovered a fossil skullcap with some ape-like features. Dubois claimed that the two fossils belonged to the same individual, and he named a new genus and species, *Pithecanthropus erectus* (the upright ape-man), the missing link between humans and their ape ancestors. Reception of Dubois's discovery was mixed, and diverse interpretations were presented – that the fossil represented an ape, a human and, following Dubois, an ape-man. The debate on the nature of *Pithecanthropus* would be resolved only decades later after many more fossil finds. It is now classified, together with other similar Asian fossils, as *Homo erectus*, a species of ancient humans from which, according to some views, modern *Homo sapiens* evolved directly.

The beginning of the twentieth century brought the most notorious affair in the history of palaeoanthropology. Amateur scientist Charles Dawson claimed to have found ancient human remains near Piltdown in Sussex, England. Soon Dawson involved the scientific expertise of Arthur Smith Woodward of the British Museum in the excavations. The search revealed several pieces of skull of the presumed human ancestor, as well as numerous prehistoric faunal remains and stone artefacts. The famous skull consisted of a human-like cranium (which testified to a large brain) and an ape-like lower jaw. The portrait of the 'First Englishman', as the specimen was sometimes called, was completed when an implement curiously reminiscent of a cricket bat was found in the same locality! Although the find stimulated numerous debates, it was generally accepted as a genuine fossil which had an important place in the human family tree.

It would turn out, forty long years later, that 'Piltdown Man' was in fact a hoax. The cranium belonged to a modern human while the fragmentary lower jaw was that of an orang-utan, and both had been stained to look older than they were. The perpetrator of the hoax is still unknown although new evidence as to his identity continues to appear. Almost anyone with an interest in human origins who inhabited the British Isles at the time has been implicated. More important than who did it, however, are the complex reasons why the scientific community was misled for such a long period of time and why no one listened to the few scientists who sensed the hoax. Piltdown's importance was supported by the most eminent scientists of the time, and to question it might have risked the academic credibility of the doubter. It is probably also significant that despite several fossils having been found in Germany, France and other European countries, none had yet been found in Britain – national pride may well have played a part in scientists' willingness to believe in 'Piltdown Man's' authenticity. There is no doubt, however, that critical to acceptance of the fossil was the way in which the Piltdown specimen seemed to corroborate the theoretical expectation of an ancestor with a large brain and ape-like jaw and teeth.

In the early twentieth century, the available fossil evidence for human ancestry and the theoretical assumptions of the day indicated that Europe and Asia were important regions for further exploration. The Central Asian plateau in particular was seen as a place which in its past had provided climatic conditions conducive to human evolution. Consequently, a series of five major palaeontological expeditions to Mongolia and China – the Central Asiatic Expeditions – took place between 1921 and 1939, led by Roy Chapman Andrews and supported by Henry Fairfield Osborn, who was president of the American Museum of Natural History. Even by today's standards, these were some of the best-financed and -supported field research excursions in history. Their aims were to survey for and collect palaeontological and archaeological materials, with the hope of uncovering clues to human origins. Although the expeditions were scientifically highly successful, no hominid fossils were recovered.

As scientific reasons seemed to point to Europe and Asia as most probable 'Cradles of Humankind', there was simply no effort to prospect for hominid fossils in Africa. Furthermore, Western imperialism created a climate in which Africa was perceived as a 'Dark Continent' inhabited by 'primitive races', from which nothing important could ever have emerged. It took several decades for shifts in both scientific theory and social context to take hold, and only then was a completely opposite position accepted – that Africa is the place from which many significant evolutionary developments emerged: the first hominids, first humans, first culture and first civilisation.

In this historical context, it is perhaps easy to see why the first hominid fossils from South Africa were both so controversial and also so critical for the development of the field – the fossils and their interpretation conflicted with all the evolutionary ideas of the time. They redirected scientific enquiry, causing huge shifts in the discipline of palaeoanthropology.

The history of South African palaeoanthropology

Fossil deposits in what was then the Transvaal province of South Africa had been documented at least as early as the 1890s, as a result of interest in calcitic speleothem cave deposits (often referred to as 'travertine' or 'lime') that were mined for use in smelting gold. In fact, the lime-mining activities by G Martinaglia at Sterkfontein were already under way in the mid-1890s, and it is likely that David Draper of the Geological Society of South Africa visited the Sterkfontein cave with other geologists in 1895 to explore for

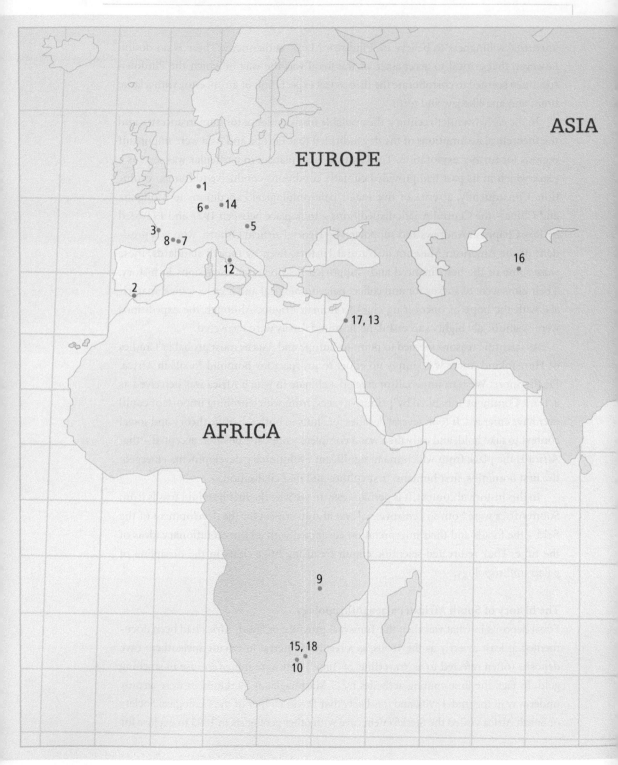

Sites of fossil excavation prior to 1947

Date	Name	Species	Place	map no.
1856	Neandertal 1	*Homo neanderthalensis*	Neander Valley, Germany	1
1864	Gibraltar 1	*Homo neanderthalensis*	Gibraltar	2
1868	Cro-Magnon 1	*Homo sapiens*	Les Eyzies, France	3
1891	Java man	*Homo erectus*	Trinil, Java, Indonesia	4
1899	Krapina C	*Homo neanderthalensis*	Krapina Cave, Croatia	5
1907	Mauer 1	*Homo heidelbergensis*	Mauer sand pits, Germany	6
1908	La Chapelle-aux-saints	*Homo neanderthalensis*	La Chapelle-aux-saints, France	7
1909	La Ferrassie 1	*Homo neanderthalensis*	La Ferrassie, France	8
1921	Broken hill 1	*Homo heidelbergensis*	Kabwe, Zambia	9
1924	Taung child	*Australopithecus africanus*	Taung, South Africa	10
1928	Peking man	*Homo erectus*	Zhoukoudien Cave, China	11
1929	Saccopastore 1	*Homo neanderthalensis*	Rome, Italy	12
1932	Skhul V	*Homo sapiens*	Skhul Cave, Mount Carmel, Israel	13
1933	Steinheim	*Homo heidelbergensis*	Steinheim, Germany	14
1938	TM 1517	*Paranthropus robustus*	Kromdraai, South Africa	15
1938	Teshik-Tash	*Homo neanderthalensis*	Teshik-tash, Uzbekistan	16
1938	Kebara 2	*Homo neanderthalensis*	Kebara Cave, Israel	17
1947	Sts 5	*Australopithecus africanus*	Sterkfontein, South Africa	18
1947	Sts 14	*Australopithecus africanus*	Sterkfontein, South Africa	18
1947	Sts 71 & 36	*Australopithecus africanus*	Sterkfontein, South Africa	18

11

4

AUSTRALIA

fossil deposits. Though it was almost another thirty years before the Taung fossil came to light, it is questionable whether any such discovery would have occurred in South Africa were it not for such lime-mining activities. (Chapters 12 and 13 discuss in more detail the role played by mining activities in spurring the development of South African palaeoanthropological science.) One can only imagine how palaeoanthropological history would be changed had hominid fossils been recovered during one of the early explorations of sites in the area in South Africa now known as the 'Cradle of Humankind'[5].

As in Europe, the first hominid remains to come to light in southern Africa represented hominids from relatively recent time periods, but to which great antiquity was attributed. Two finds in particular are significant – the 1913 partial cranium from Boskop (Potchefstroom District, Transvaal – now North West province) in South Africa, and the 1921 cranium from Broken Hill (Kabwe), Zambia. These specimens were recognisably different from each other, but both were large-brained with robust, heavy brow ridges, large faces and (for Boskop) thick cranial bone – features immediately inviting comparison with Neandertals and even with *H. erectus*. These fossils were isolated finds – Boskop without any associated artefacts or faunal material, and Broken Hill during mining operations – but were attributed generally to the Middle Stone Age. In the then current context of views invoking cultural diffusion from European populations, these fossils were generally thought to represent one of many ancient 'types' or 'races' that had populated southern Africa during the Middle Stone Age. Boskop, Broken Hill and later finds were repeatedly discussed amid long-term controversies concerning the origin of racial 'types' such as the 'Bushmen', 'Hottentots' and 'Bantu', well into the 1960s.

The story of Raymond Dart's involvement in South African palaeoanthropology has been told many times in great detail. Briefly, Dart was born and received his medical training in Australia, and worked for a time at University College, London, under the famous British anatomist Grafton Elliot Smith. Dart arrived in Johannesburg in early 1923 from London to assume the position of Professor of Anatomy at the relatively new University of the Witwatersrand Medical School. It is of interest to note that the principal of the university expressed his regret at the appointment of an Australian, reflecting the extent of social and political bias at the time.

Dart approached his new post with characteristic vigour. In order to build up a teaching collection for the ill-provisioned department, he challenged students to bring in anatomical and skeletal specimens of interest. One student, Josephine Salmons, brought in a fossilised baboon skull she had obtained through a friend of her family who was a director of the Northern Lime Company. At the time, a fossil baboon skull in southern Africa was notable enough in itself, and Dart consulted a geology colleague at

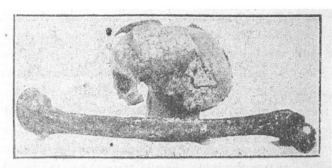

SKULL AND THIGH BONE FROM TZITZIKAMA.
The discovery of these remains in the coastal area has confirmed the indications given by the Boskop skull found in the Transvaal that a remarkable and hitherto unknown type of early man lived in South Africa.

Fig. 2.5 The Boskop partial cranium

Fig. 2.6 Taung site

the university, RB Young, to obtain further information about the lime-mining operations from which the baboon skull had been recovered – a site known as Taung, in what was then the Bechuanaland Protectorate, near Kimberley (today part of North West province). As it happened, Young was a geological consultant for the lime-mining industry, and was already familiar with the Taung lime works. Thus he was able to arrange for crates of the conglomerate fossil-bearing rocks (which are called 'breccia') from Taung to be sent to Raymond Dart in Johannesburg.

Two crates of fossil breccia from Taung arrived on Dart's doorstep in late November 1924, one of which contained the fossilised braincase and facial skeleton of what came to be known as the 'Taung child': a juvenile individual representing an early human ancestor that Dart named *Australopithecus africanus* – the 'southern ape of Africa' (see pages 40–41). His description of the Taung fossil documented a number of human-like features of the brain, facial skeleton and dentition that indicated to Dart that he had identified the first 'man-like ape', an ancestral species between modern apes and humans, but definitely on the lineage of human ancestry. *A. africanus* and related hominids from South and East Africa came to be known as the australopithecines.

Raymond Dart's announcement and description of the 'Taung child' fossil was published in the journal *Nature* in February 1925. If Dart's ideas were correct, then human origins were to be found in Africa, not Europe or Asia, and furthermore, a large brain was not the first feature to evolve in our ancestors. The story of the subsequent controversy over, and eventual acceptance of, his ideas is a long and protracted one, but the important point is that Dart's interpretation of this single, unexpected fossil specimen fairly literally turned the palaeoanthropological world upside down, and caused widespread reaction in academia, high society, and even political circles. The discovery eventually caused a revolution in thinking – a paradigm shift – about human origins and prehistory, and serves well as the event leading the field of palaeoanthropology to its modern state.

But first, Dart and his supporters needed to find further proof in order to counter the many criticisms published by the eminent British researchers Sir Arthur Keith, Sir Arthur Smith Woodward and Dart's own mentor, Sir Grafton Elliot Smith – as well as by other prominent scientists. The initial objections to Dart's interpretation of the Taung fossil resulted from several factors. Firstly, the fossil demonstrated morphological characteristics that were intermediate between modern apes and humans, and thus it was strongly argued by

Fig. 2.7 The issue of Nature *of 7 February 1925, Vol. 115 No. 2884, in which Dart published his announcement and description of the 'Taung child'*

NATURE   181

SATURDAY, FEBRUARY 7, 1925.

CONTENTS.
                                                                   PAGE
The Future of the British Patent Office . . . 181
The Imperial College of Tropical Agriculture . 183
Reminiscences of Great Naturalists. By Prof. J.
   Arthur Thomson 184
Modern Views on Cytology. By Prof. J. Brontë
   Gatenby 185
General Chemistry 187
Our Bookshelf 188
Letters to the Editor:
   The Origin of Sponge-Spicules.—Prof. Arthur
      Dendy, F.R.S. 190
   On the Excitation of Spark Spectra.—Sven Werner 191
   Rainfall Correlations in Trinidad.—W. R. Dunlop 192
   Astrophysics without Mathematics.—Prof. Herbert
      Dingle; Prof. E. A. Milne 193
   The Structure of the so-called Ultraviolet Bands of
      Water Vapour.—G. H. Dieke 194
   Hafnium Oxide in Tungsten Filaments.—J. A. M.
      van Liempt 194
   Citrus Fruit and Scurvy.—Prof. W. A. Osborne . 194
Australopithecus africanus: The Man-Ape of
   South Africa. By Prof. Raymond A. Dart . 195
Biographical Byways. By Sir Arthur Schuster,
   F.R.S.—6. S. P. Langley 199
Obituary:
   Dr. J. M. Ellis McTaggart. By Prof. G.
      Dawes Hicks 199
   Mr. C. H. Wordingham. By A. R. . . . 200
   Mr. George Abbott 201
Current Topics and Events 201
Our Astronomical Column 205
Research Items 206
Scientific Work of the Fishery Board for Scotland.
   By Prof. W. C. McIntosh, F.R.S. . . . 209
Science and the Instrument Industry . . . 209
The Botanic Garden, Copenhagen . . . 210
University and Educational Intelligence . . . 211
Early Science at Oxford 212
Societies and Academies 212
Official Publications Received 213
Diary of Societies 215

Editorial and Publishing Offices:
MACMILLAN & CO., LTD.,
ST. MARTIN'S STREET, LONDON, W.C.2.

Editorial communications should be addressed to the Editor.
Advertisements and business letters to the Publishers.

Telephone Number: GERRARD 8830.
Telegraphic Address: PHUSIS, WESTRAND, LONDON.

NO. 2884, VOL. 115]

The Future of the British Patent Office.

THE British patent system is a matter which concerns all workers in applied science, for it represents an attempt—faulty and incomplete, but still an attempt—to secure for such workers the credit for their achievements, together with a share of the material advantages arising from these. Hence any event which seriously affects the future of the patent system is one to which the scientific world should give careful consideration, and such an event is just beginning to appear on the horizon. Lest it should take shape before its implications have been seriously canvassed, it may be well to direct attention to some of its aspects. There is a rule which requires Government servants to submit to superannuation at an age when many men are still capable of their best work, and since the rule appears to be inexorably applied, the retirement of the present Comptroller of the Patent Office and the appointment of his successor must be regarded as inevitable in the not very distant future. It is perhaps a little early to discuss this question, but not too early; for when the first official intimation of such a change is given, the selection of the successor may be actually, if not formally, a *fait accompli*.

That a scientific office should have a scientific man at its head is a principle which seems obvious but needs to be constantly reasserted, because the administrative officials who influence such appointments are not always sympathetic towards the claims of science. In fact, a lack of sympathy in that direction is sometimes manifested to a degree which exposes it to strong criticism, as in the proceedings of committees A, B, and C of the National Whitley Council, and in the general tendency to regard the man of science as a mere adviser who is himself incapable of administrative work. Such an attitude is the more unjustifiable from the fact that a scientific training is necessarily always *additional* to some degree of education in the humanities, whereas a literary scholar may be quite ignorant of science; so that the former type of upbringing is the more likely to produce the breadth of outlook which is necessary in handling men and affairs.

Taking the British Patent Office as an example, let us examine the qualifications which are necessary in the man who is to direct its labours. The duties of the Comptroller fall under three heads as follows:

(1) He is the senior Hearing Officer for disputes as to patents, trade marks, and designs. He has to adjudicate in "oppositions" brought by interested parties against the grant of particular patents, as well as in cases where examiner and applicant fail to agree in regard to the official requirements put forward by

these researchers on the same evidence that it was a fossil ape, not a fossil human. Secondly, the small size of the braincase ran counter to prevailing views, and the incompleteness of the specimen left many details of cranial form unspecified. Thirdly, the fact that the 'Taung child' was a juvenile meant that it was impossible to compare its morphology with relevant adult fossil specimens. Finally, in the absence of good geological dating techniques, it was also thought by some that the Taung fossil was much too young to have anything to do with human ancestry – yet ironically, it was the oldest recovered evidence for hominid evolution at the time!

After the Taung discovery, Dart received occasional crates of fossil materials from the lime workings at Sterkfontein and other sites. But no additional hominids were recovered, and no concerted efforts to excavate any such sites were made. However, a science schoolteacher from Pietersburg (now Polokwane), WI Eitzman, visited the lime works in the Makapansgat Valley (in the Northern Transvaal, now Limpopo province) in 1925 and arranged for crates of fossils to be sent to Dart. These fossils consisted mostly of bovid (antelope) bones, and Dart noted that many of them were covered by a blackened stain. Dart organised chemical analysis of the fossils, which indicated the presence of isolated carbon; this suggested to him that the bones had been burnt prior to burial and fossilisation. He concluded that ancestral hominids had cooked and burnt such bones, and this idea seems to have been the initial stimulus for his later and more elaborate osteodontokeratic (ODK) hypothesis for australopithecine cultural behaviour[8].

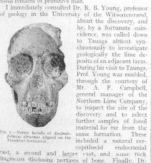

RAYMOND DART AND THE 'TAUNG CHILD'

Fig. 2.8 Front and oblique views of the 'Taung child'

When Raymond Dart published his announcement about the 'Taung child' in 1925, and established *Australopithecus africanus* as an ancestral 'missing link' to humans, he made suggestions that directly conflicted with established views about the pattern of events in human evolution. Perhaps the most challenging issue among his claims was that a small-brained, yet already bipedal species of hominid had a position of significance in the human evolutionary lineage. Conventional views at the time claimed that large brains were a key evolutionary feature identifying human, as opposed to ape, ancestry, and that other evolutionary features only appeared after the advent of a large brain.

However, there were other features of the 'Taung child's' morphology that were both important and interesting, and Dart's entire scenario for the fossil was based on his insightful, if not speculative, interpretation of the anatomical features of this specimen.

One of the first things Dart noticed was that the brain size indicated by the fossil was too large to represent a fossil monkey such as a baboon; this suggested immediately that the fossil was of an unknown form. However, Dart was extremely excited about the morphology he observed on the endocast, the fossilised impression of the inner surface of the braincase itself, which can preserve details of the convoluted gyri (elevations) and sulci (grooves) of the brain's cerebral cortex as well as other details. Dart had trained as a neuroanatomist under the famous British anatomist Grafton Elliot Smith, and he immediately noticed certain features on the endocast of the Taung fossil that indicated human-like, or 'advanced', brain function in comparison to apes, despite the smaller size of the brain.

Secondly, Dart noted that the canine tooth crown of the Taung individual was smaller and less projecting than that expected for an ape, even considering its juvenile status. Only much later, in 1929, was Dart able to separate the lower jaw from the cranium and

Fig. 2.9 Raymond Dart with australopithecine reconstructions in 1955

observe the tooth crowns of the molars and premolars; it was then possible to confirm that the fossil was certainly not an ape, and was more closely related to humans. Other aspects of the face and jaw of the specimen were also more human-like, in particular the degree of prognathism, or facial projection – the 'Taung child' did not have a large projecting face as would be expected in an ape.

Finally, Dart noted that the *foramen magnum* (Latin for 'large opening'), through which the spinal cord leaves the braincase to enter the vertebral column, was positioned relatively far forward on the base of the skull. This indicated that the head was balanced atop an essentially vertical trunk – evidence that Taung walked with an erect posture, and was thus likely to have been a biped.

Dart's description of the 'Taung child' specimen laid out a series of morphological features for a group of early hominids that became known as the australopithecines – members of the earliest-known family of human ancestors. These included bipedality, small brains, and small, stout canine teeth, as well as other features of the facial skeleton and brain endocast. Dart's identification of this group of characteristics amounted to a bold suggestion that the evolutionary transition from 'ape' to 'human' had taken place in the absence of a large brain, and was in direct conflict with the prevailing theories of the day.

Although a few scientists had been supportive of Dart's claims regarding the Taung fossil, it was clear that the matter would not be resolved without analysis of additional fossils. That cause was taken up by Dart's most ardent supporter, an already well-established palaeontologist, Dr Robert Broom. The enthusiasm and intensity that Broom brought to this task is clearly expressed in Dart and Craig's description of Broom's first encounter (in February 1925) with the 'Taung child' fossil:

> Broom immediately wrote a letter of congratulations and two weeks later burst into my laboratory unannounced. Ignoring me and my staff, he strode over to the bench on which the skull reposed and dropped on his knees 'in adoration of our ancestor,' as he put it…Having satisfied himself that my claims were correct, he never wavered (Dart & Craig 1959: 35).

Robert Broom was a charismatic and enigmatic character (see pages 44–45) and a prolific scientific researcher. For most of his scientific career, he made his living as a 'country doctor' and engaged in his palaeontological and scientific pursuits on his own accord. It was significant for his career, and therefore for South African palaeontology as a whole, that General JC Smuts was also an avid supporter of science in general and of palaeontology in particular. He was on personal terms with Dart, Broom and other South African scientists, and did much to encourage and support palaeoanthropology during this period. Smuts ended his first term as South Africa's prime minister at about the time that Dart announced the 'Taung child', and he became the president of the South African Association for the Advancement of Science in 1925. Smuts helped to arrange a permanent position for Broom at the Transvaal Museum in Pretoria beginning in 1934, an appointment which allowed Broom to devote his attention fully to palaeontology. Though he was 67 years old at the time, Broom immediately set to work exploring cave sites in what is now the Cradle area, and also further north in the Makapansgat Valley. Within two years he had recovered the first adult australopithecine from the Sterkfontein lime works.

Between 1936 and 1938, Broom recovered more *Australopithecus africanus* fossils (initially named *Australopithecus transvaalensis*) from Sterkfontein, and notably the first adult australopithecine specimens. Though the lime works was closed in 1939, and the outbreak of World War Two further disrupted research, Broom continued to work at Sterkfontein intermittently through the 1940s. This work resulted in the recovery of several significant new *A. africanus* fossils, including the nearly complete cranium (Sts 5[9] or 'Mrs Ples') and an associated partial skeleton (Sts 14) in 1947.

In addition to his work at Sterkfontein, Broom recovered the first fossil hominids from the nearby sites of Kromdraai and Swartkrans. In 1938, following a find by schoolboy Gert

Terblanche, Broom identified fossils from Kromdraai as a new hominid species, *Paranthropus robustus*. These fossils demonstrated many unique characteristics compared to the Sterkfontein specimens, such as enlarged cheek teeth (molars and premolars), small incisors and canines, and heavy markings for attachment of chewing muscles. Broom initially considered that these specimens represented a species more closely related to humans, though they are now recognised as an extinct evolutionary 'side-branch' of hominids.

The first major monograph on the South African australopithecine material was published in 1946 by Broom and GWH Schepers, a Professor of Anatomy on Dart's staff, as the *Transvaal Museum Memoir No. 2*. Published with financial support organised by JC Smuts, this monograph described and summarised all the material so far recovered from Sterkfontein, Kromdraai and Taung, and was such a convincing presentation of the available material that even Sir Arthur Keith was compelled to change his opinion on the status of the australopithecines.

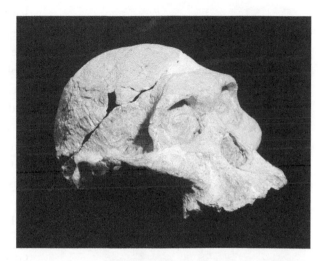

Fig. 2.10a Sts 5, 'Mrs Ples'

Another great personality in South African palaeontology was James Kitching. In 1947, Kitching and his brothers were assisting with the long-term excavations at Cave of Hearths in the Makapansgat Valley, directed by CR van Riet Lowe and then by Revil Mason. On a weekend excursion to the lime-works dumps, Kitching recovered a fossilised australopithecine occipital fragment from Makapansgat. This find, along with later fossils recovered during student trips by the young

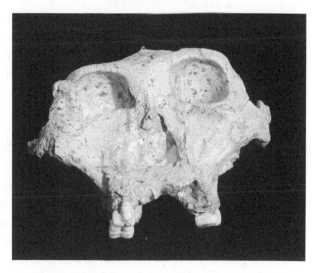

Fig. 2.10b SK 48, Paranthropus robustus *from Swartkrans*

ROBERT BROOM

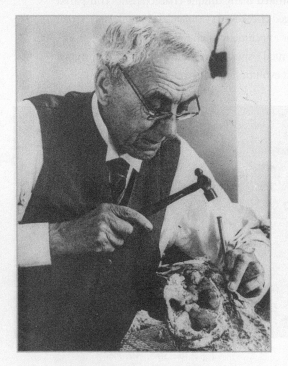

Fig. 2.11 Robert Broom at work

Robert Broom (1866–1951) was born in Paisley, Scotland. He obtained his medical degree at Glasgow University in 1889. In the early 1890s his adventurous spirit took him twice to America and later to Australia where he practised medicine and conducted scientific research. In 1897 he arrived in South Africa where, with only brief interruptions, he remained until the end of his life. Just as he had in Australia, he worked as a medical practitioner and conducted scientific research in his spare time, mostly on fossil reptiles. In 1903 he was appointed Professor of Geology and Zoology at the Victoria College (later the University of Stellenbosch). He returned to medical practice and the life of a freelance scientist six years later. In 1929 he retired (unsuccessfully) from his medical practice and moved to Grahamstown, trying to dedicate all his time to scientific research. In 1934 he was appointed Curator of Palaeontology at the Transvaal Museum. He stayed at this institution until the end of his life, which in Broom's case coincides with the end of his scientific career as he wrote the final sentences of his last monograph virtually on his deathbed. Because of his boastful personality and strange habits, such as belief in the power of the sun, which sometimes led him to excavate naked, Broom is often described as an eccentric. This eccentricity in the form of inclination towards heterodoxy is felt even in his scientific output.

Broom's early scientific career was devoted to comparative anatomy. These early studies were followed by research in palaeontology, physical anthropology and palaeoanthropology. Later in life Broom's interest in the meaning of evolution led him to philosophical speculations. In palaeontology he is best known for the numerous discoveries of fossil reptiles that he made in the Karoo region of South Africa. His theory of evolution of mammals from mammal-like reptiles had enormous impact and was crucial in bringing him the Fellowship of the Royal Society, London, in 1920.

He also developed an interest in local populations of southern Africa and published several influential papers on the topic. While his research, based on the typological race concept, was acceptable in the early twentieth century when his work was published, his ways of acquiring human skeletal material, which included grave-digging and boiling of human corpses, were almost as outrageous in his time as they are today.

Broom was one of the first scientists (in 1925) to support Dart's interpretation of the newly discovered *Australopithecus africanus* as a possible ancestor of later hominids, including modern humans. However, his more intensive involvement in palaeoanthropology would start after his move to the Transvaal Museum. In 1936 he recovered australopithecine remains from the site of Sterkfontein. These were the first in the series of several major palaeoanthropological discoveries which he continued to make until his last days at the Sterkfontein, Kromdraai and Swartkrans sites. These finds include the 1947 discovery of 'Mrs Ples', one of the symbols of South African palaeoanthropology. At the end of the Second World War he published the classical monograph *The South African Fossil Ape-men: The Australopithecine* with the young neuroanatomist GWH Schepers, which induced a paradigmatical change in the understanding of human evolution.

A materialist, atheist and convinced evolutionist as a student, Broom would soon change his outlook (partly as a result of the influence of spiritualism on him) and try to create a synthesis of religion and science. He firmly believed in evolution but not in Darwinism, and claimed that spiritual forces lay behind the evolutionary processes.

Robert Broom is one of the most significant and most interesting figures in South African science. He is usually presented in an old-fashioned 'great hero of science' manner. It should not be forgotten, however, that many of Broom's ideas are quite foreign to modern science while some, deeply embedded in cultural values of the era to which he belonged, are today regarded as outrageous. His scientific career must be understood in its entirety, as it exemplifies how South African palaeoanthropology and science in general were practised in the first half of the twentieth century.

PV Tobias, stimulated Raymond Dart to return to palaeoanthropology, particularly to the development of his ODK cultural model. Research at Makapansgat continued intermittently through the 1960s and 1970s under Dart, and later Tobias, both ably assisted by Alun R Hughes.

In 1948, Broom and his young protégé JT Robinson opened up a new excavation at Swartkrans, a hilltop site near Sterkfontein. International finance and assistance for this work were provided by Wendell Phillips of the University of California's Africa Expedition, and for a short time excavations proceeded jointly at Sterkfontein and Swartkrans. The Americans were only involved for a few months, but the new excavations at Swartkrans proved to be extremely productive, and continued until 1953. In addition to new *P. robustus* material (which Broom called *P. crassidens*), a novel species of early hominid was recovered and named *Telanthropus capensis* (this specimen is now known as *Homo ergaster*, an African form of *H. erectus*). This find was the first evidence suggesting that two extinct early hominid species had once coexisted. In addition, some of the most complete *Paranthropus*[10] material was recovered in work done during this period, including the cranium SK 48 and the mandibles SK 12 and SK 23. Work at Swartkrans ceased after this period and did not recommence until 1965, when CK Brain initiated new excavations there.

During the period from 1953 to 1965, a number of important events occurred both in South Africa and internationally. Firstly, the work of Joseph Weiner and Kenneth

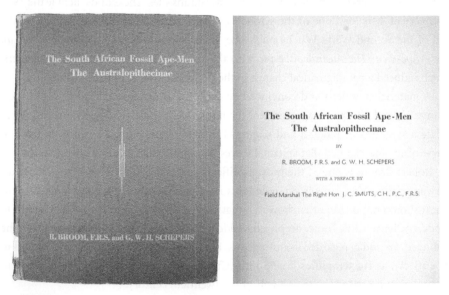

Fig. 2.12 *The cover and title page of Broom and Schepers's monograph published in 1946 as* Transvaal Museum Memoir

Oakley from 1953–1955 exposed the Piltdown fossil as a hoax, and finally alleviated the confusion caused by this big-brained specimen. In South Africa, additional research on various aspects of the fossil fauna and geology of the sites in the Cradle area was conducted by the Transvaal Museum, largely under the direction of CK Brain. A number of important research monographs were published by the Transvaal Museum, including Broom's monograph on *Paranthropus* (1952 – posthumously, after his death in 1951), Robinson's monograph on australopithecine dentition (1956), and Brain's volume on the geology of the australopithecine-bearing cave deposits of the Transvaal (1958). These publications provided a wealth of documentation for the international community, and further confirmed the status and significance of the South African fossils for an understanding of human evolution.

It was also during this period that Louis and Mary Leakey announced the finding of the first early hominid fossils from Olduvai Gorge in Tanzania. The OH 5 robust australopithecine cranium was recovered in 1959 and named *Zinjanthropus* (now *Paranthropus*) *boisei*. In 1964, a new species named *Homo habilis* was announced, again providing evidence for the contemporaneity of different early hominid species. These East African fossils were extremely important for several reasons. Firstly, these finds documented the presence of australopithecines and later hominids in East Africa, and demonstrated that they were widespread and not just an isolated phenomenon in South Africa. Secondly, the East African material was found in association with primitive stone tools, indicating that at least one of these early ancestral hominids was capable of clearly human-like behaviour. Thirdly, the newly-developed potassium-argon dating technique was utilised in 1961 to produce an absolute age estimate of 1.75 million years ago for the OH 5 cranium. Previous age estimates for such early fossils were both relative and theoretical, and a period of about 1 million years was allowed for the entire period of human evolution. Thus, the OH 5 date almost doubled the timescale for hominid evolution! Since then, even earlier hominid fossils have been recovered in Africa and elsewhere and dated with such techniques. Absolute dating methods have had a dramatic effect on interpretations of the hominid fossil record, simply by providing an accurate chronological scale for the events of hominid evolution.

The period after 1965 was one of modernisation of excavation techniques and of more rigorous and scientific interpretations of the hominid fossils and other faunal material. Brain initiated long-term excavations at Swartkrans, and in 1966 Tobias and Hughes began excavations at Sterkfontein that are still in operation to the present day under the direction of RJ Clarke. These excavations have recovered literally hundreds of new hominid fossils and hundreds of thousands of associated faunal remains from these sites. Brain's work at Swartkrans is recognised as the first rigorous application of taphonomy, or the

Fig. 2.13 Ron Clarke, Stephen Motsumi and Nkwane Molefe

science of burial processes, in palaeoanthropology, and his interpretations of the fossil bone assemblages conclusively countered Dart's earlier and more imaginative ODK model, and allowed for more realistic and inform-ative interpretations of early hominid behaviour. The work at Sterkfontein has resulted in the recovery of over 600 hominid specimens – predomi-nantly of *Australopithecus africanus*, but including specimens thought to represent *H. habilis* and perhaps even some robust australopithecines. In addition, the most recent spectacular finds at Sterkfontein include the essentially complete *Australopithecus* skeleton recovered by Ron Clarke, Stephen Motsumi and Nkwane Molefe between 1994 and 1997. This find, and more material announced in 2003, were initially assigned dates older than the *A. africanus* material from Sterkfontein; however, recent re-dating has produced controversial younger dates for the same deposits.

The impact of South African palaeoanthropology on our understanding of human evolution

Raymond Dart's initial discovery at Taung had an immediate worldwide impact in the field of palaeoanthropology – how different it would be with-out the South African australopithecines! His claims were controversial, and required that further exploration be undertaken to recover additional fossils. Thus, the work at the sites in what is now known as the 'Cradle of Humankind', involving the efforts of Robert Broom and many others, was critical in the development of our current understanding of our evolution-ary heritage – something which applies to every human being alive today.

Because of the significance attributed to these finds, both by other South African researchers and by influential individuals such as JC Smuts, South Africa attained recognition as a world centre for palaeoanthropological research, and to this day produces important new discoveries of fossils and ground-breaking research. It is certainly paradoxical that many of the personalities involved held views about human beings that are divergent from modern views of humanity, but that historical pathway was necessary to develop our modern understanding. The importance of research into our common heritage and evolutionary past, and the recognition it brings, is still reflected in the investment that the current South African government has made in preserving and developing the sites and fossils of the Cradle and elsewhere.

Fig. 2.14 The excavation site at Sterkfontein

CHAPTER 3

FOSSIL HOMINIDS
OF THE 'CRADLE OF HUMANKIND'

Kevin Kuykendall

The Plio-Pleistocene fossil sites of the 'Cradle of Humankind' in South Africa have produced an important and extensive collection of fossil hominids – a group of primates including modern humans, their direct ancestors, and closely related species. However, the history of research on human evolution does not begin in South Africa, and the first fossils recognised as human ancestors came from Europe and Asia, representing relatively late periods in human evolutionary history.

At the time that Charles Darwin published *On the Origin of Species* in 1859 there was barely any known hominid fossil record, and the few fossils that had been recovered during his lifetime stimulated so much controversy that their true significance was not yet realised at the time of his death in 1882. However, Darwin made two speculations that contributed significantly to views about human evolution in the late nineteenth and early twentieth centuries. Firstly, he claimed that on the basis of the geographic distribution of living great apes (chimpanzees, gorillas, and orang-utans – even then recognised as our closest living evolutionary relatives), we should expect to find ancestral human fossils in either Africa or Asia. Second was his prediction that the human ancestor would be intermediate between modern humans and apes, exhibiting many ape-like characteristics. However, Darwin also suggested that a superior intellect, accompanied by a requisite large brain, would have been necessary even in early ancestors in order for 'human-like' traits to have evolved.

The first claim was cited by Dubois in justifying his choice to excavate in Asia (at Java, Indonesia, in the 1890s) where he discovered hominid fossils now attributed to *Homo erectus*. It was also used by Dart in 1925 to bolster support for his interpretation of the 'Taung child', for which he named the taxon *Australopithecus africanus* from South Africa as an early and primitive 'missing link'. (These finds are discussed more fully in Chapter 2.)

However, it turned out that part of Darwin's second prediction was not correct. While the earlier finds of Neandertal and *H. erectus* fossils seemed to support the idea of large-brained human ancestors, they did not demonstrate many ape-like characteristics. In addition, the unexpectedly small brain and additional ape-like features of *A. africanus* were certainly in conflict with prevailing ideas about the pattern of human evolution, even after consideration of the postcranial bipedal features that these fossils demonstrated.

The recovery and recognition of the South African australopithecine fossils from sites such as Taung, Sterkfontein, Kromdraai and Swartkrans was very important for the development of our modern understanding about the human evolutionary past simply because they were so unexpected – their description, assessment and eventual acceptance in the mid-twentieth century brought about a revision of ideas concerning human evolution. While Chapter 2 focused on the development of the modern scientific discipline of palaeoanthropology, this chapter will describe and discuss these hominid fossils, and their significance in the broader scheme of palaeoanthropological research.

Early hominid evolution

The taxonomic family of primates that includes living humans, their direct ancestors and close relatives is known as the *Hominidae*, or hominids. (An alternative terminology, based largely on genetic analyses, is *Homininae* and hominins.) Hominids are distinct from modern apes, their ancestors and close relatives because of the fact that they are on the evolutionary pathway, or lineage, leading eventually to modern humans, and thus share certain distinguishing morphological traits (see Figs 3.1 and 3.2). However, evolutionary theory (and the fossil record) suggests that the further back in time we explore, the greater the similarity we should expect to see between human and ape ancestors, until both lineages are traced to a common ancestor – a single species from which both modern humans and chimpanzees are ultimately derived. Theoretically, and considering both genetic evidence and the fossil record, it would be difficult to distinguish clearly between fossils of ancestral hominids and chimpanzees at, say, 6 million years ago – that is, soon after divergence from the common ancestor. It is thus very interesting that until recently, palaeoanthropologists did not recognise any fossil African apes, though new and more primitive fossils attributed to early hominids are recovered every few years. A number of discoveries since the mid-1990s have resulted in the naming of new hominid species and even genera, such as *Sahelanthropus* (6–7 million years ago), *Orrorin* (6 million years ago), *Ardipithecus* (4.4–5.8 million years ago), and *Kenyanthropus* (2–3 million years ago), in addition to new species of *Australopithecus* and *Homo*. And then in 2005, Sally McBrearty and Nina Joblonski recovered two maxillary incisors and one maxillary molar from the Kapthurin Formation in Kenya representing a fossil chimpanzee dated to about 500 000 years ago. These are the first fossils relating unequivocally to modern African apes.

The ancestral species on the lineage to modern humans are frequently referred to in the popular media as 'missing links' – intermediate forms that are somehow 'in between' modern apes (specifically chimpanzees) and humans. This popular concept is erroneous for several reasons. Firstly, evolutionary change is continuous (though not necessarily gradual) and it is often difficult to define discrete evolutionary 'units' – fossil specimens vary, and may demonstrate morphological gradations between established species that do not fit neatly into recognised taxa. Secondly, the fossil record indicates that there were often two or more hominid species existing contemporaneously – and not all such extinct taxa can be directly ancestral to later species. Thirdly, the only taxon that would approximate the idea of an intermediate 'missing link' between modern chimpanzees and humans is their common ancestor itself. Any hominid fossils are by definition part of the ancestral lineage between that common ancestor and our own modern species, and have nothing to do with chimpanzee evolution. So far, that common ancestor has eluded palaeoanthropological discovery, thus the 'true missing link' remains unknown. The reality is that though we currently identify as many as fifteen or twenty hominid species, it is simply impossible on existing evidence to define a conclusive direct lineage of 'human ancestors', and some of the extinct species are not 'links' at all!

However, at the time of their recovery and recognition in 1925, when the known hominid fossil record was extremely sparse, the South African australopithecines certainly did fit the concept of an intermediate form linking modern humans and apes. The fossils demonstrated some seemingly ape-like traits, such as small brains and large,

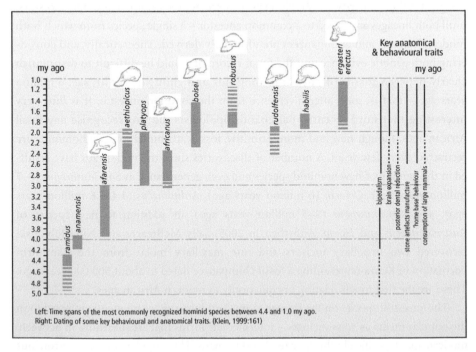

Left: Time spans of the most commonly recognized hominid species between 4.4 and 1.0 my ago.
Right: Dating of some key behavioural and anatomical traits. (Klein, 1999:161)

Fig.3.1 Temporal distribution of different hominid species between 1.0 and 4.4 MYA

relatively forward-projecting (or prognathic) faces. But they also possessed human-like traits of the dentition such as reduced canine teeth, more-or-less bicuspid premolars, and thick enamelled tooth crowns. Finally, the australopithecine fossils included fragmentary postcrania – bones of the trunk and limbs – that clearly demonstrated features associated with bipedal locomotion – a notably human adaptation.

Thus, a number of evolutionary trends can be identified from the hominid fossil record. Relating to the skull and dentition, the first is an increase in cranial capacity (cubic centimetres, or measure of brain size) from approximately 400–500 cc in australopithecines to about 1 350 cc on average in modern humans. The second evolutionary trend involves changes to the dentition, including reduction of the canine tooth in more recent forms, as well as changes in cheek tooth size. However, this involves both an early increase in premolar and molar tooth size (called *megadontia*) in the genera *Australopithecus* and especially in *Paranthropus*, as well as a later decrease in overall tooth size in the genus *Homo*. A third trend, associated with this decrease in tooth size and in the entire masticatory complex (chewing structures such as the jaws, teeth, and their associated muscles), is the general reduction in the size and prognathism of the facial skeleton – in apes and early hominids, the jaws 'jut forward' while in modern humans the face, teeth, and jaws are orientated under the braincase, and thus our facial profile is more vertical.

Evolutionary trends of the postcranial skeleton (that is, below the cranium, or of the body rather than the head) generally involve bipedalism and body size. The fossil record indicates that all hominids were to some degree bipedal – they walked upright on two feet like we do. However, limb proportions were different in earlier hominids, in that they had longer upper limbs and phalanges (fingers) that would have been adapted for climbing in trees. In addition, the earlier hominids were much smaller than we are, weighing in at about 30–40 kilograms, and standing between 1.1–1.4 metres tall, with males somewhat heavier and taller than females (such differences between male and female form are called *sexual dimorphism*). Thus, a final evolutionary trend is that hominid bipedalism became more fully terrestrial, and that body size increased over time.

We now know that a number of species of australopithecines existed and shared many of these general features, thus palaeoanthropologists have had to identify other traits to distinguish among different early hominid species. Our interpretations regarding the type of animal represented by such fossils have likewise changed as more and more new fossils have come to light. It is fortunate that the hominid fossil record has become so abundant, but new fossils often result in as many questions as answers.

In fact, the initial recovery of the 'Taung child', followed by other South African australopithecine material, raised so many questions and the interpretations given differed so much from expectations of the times, that they raised a scientific controversy that lasted thirty years. But when the original interpretations of the fossils were finally accepted, a revolution in thinking about the human evolutionary past had occurred. The hominid fossils from the Cradle and other South African sites played a crucial role in confirming these 'revolutionary–evolutionary' ideas.

Following current terminology, South African early hominids are placed into three groups that correspond well with taxonomy at the genus level, though the evolutionary relationships among them remain debatable (see Figs 3.1 and 3.2). First are the gracile australopithecines, placed in the genus *Australopithecus* and represented by well-known specimens such as the 'Taung child' and Sts 5 or 'Mrs Ples' from Sterkfontein, both attributed to the species *A. africanus*. The fossils from Makapansgat are also attributed to this species, as well as a few isolated teeth and fragments from other sites in the Cradle. Finally, the associated skeleton and skull from Member 2 at Sterkfontein, Stw 573 or 'Little Foot', whose discovery took place between 1994 and 1997, has been identified by Ron Clarke as an australopithecine, but not *A. africanus* because of some primitive features observed. In East Africa, the closest hominid relative to *A. africanus* is *A. afarensis*, represented by the famous 'Lucy' partial skeleton and additional material from the Hadar region in Ethiopia and Laetoli, Tanzania.

The second group is generally referred to as the robust australopithecines, which are placed in the genus *Paranthropus*, and include fossils from Swartkrans (such as the

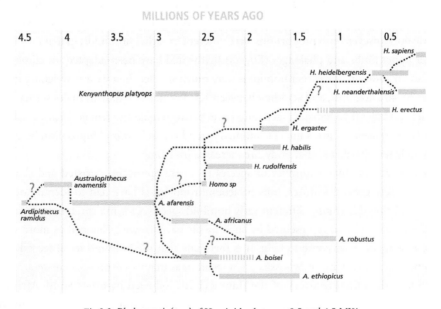

Fig.3.2 *Phylogenetic 'tree' of Hominidae between 0.5 and 4.5 MYA*

SK 48 cranium and the SK 12 mandible) and Kromdraai (including the type specimen TM 1517), as well as the more recently recovered fossil assemblage from Drimolen. There is also an isolated molar tooth from Gondolin, and some researchers would also place a few of the larger isolated molars from Sterkfontein in this group. Robust australopithecines in South Africa are called *Paranthropus robustus*, but the Swartkrans material has sometimes been referred to as a distinct species, *P. crassidens,* as originally named by Robert Broom. In contrast, the East African robust hominid material from sites such as Olduvai Gorge, Koobi Fora, and the Omo is generally classified as a distinct species, *Paranthropus boisei,* and some researchers recognise yet another and earlier East African species, *P. aethiopicus.* The dates from the South African robust australopithecine sites are generally between 1–2 million years ago, and some sites in East Africa are dated earlier. Thus, robust australopithecine species are younger than those of gracile forms in both regions, and are considered to be evolutionary descendants.

The terms 'gracile' and 'robust' are currently used primarily to describe the differences in the dentition and masticatory (chewing) structures of the skull in these species. The robust forms have been described as 'chewing machines', with larger teeth, more massive faces, and other structural features that relate to a reliance on a diet requiring heavy chewing. These specialised masticatory structures are thought to have evolved as an adaptation to a diet consisting of coarser, lower-quality foodstuffs such as roots, tubers, and perhaps thick-husked fruits. In contrast, the gracile australopithecine diet is thought to have included more varied and higher-quality food items such as fruits, leaves and animal protein. Additional information on early hominid dietary interpretations will be presented later.

Such variations in anatomy and structure indicate that the two early hominid groups were adapted to different dietary niches,[1] and many researchers prefer to place them in two separate genera, *Australopithecus* and *Paranthropus,* as I do in this chapter. This adaptive difference is also reflected in differences in the pattern and timing of dental development, indicating that *Australopithecus* and *Paranthropus* juveniles grew up differently. Other researchers focus on the similarities in brain size, postcranial anatomy, and the comparatively large tooth size of gracile and robust australopithecines in comparison to the genus *Homo*, and place both taxa in the genus *Australopithecus.* The difference is that classifications placing each group in a separate genus emphasise that gracile (*Australopithecus*) and robust (*Paranthropus*) australopithecines had evolved distinct adaptive strategies which are reflected in their morphological and developmental differences.

The third major group of early hominids whose remains have been recovered at sites in the 'Cradle of Humankind' includes early species in our own genus, *Homo*. Fossils that have been attributed to *Homo habilis* have been recovered from Sterkfontein, and some designated as *Homo ergaster* have been identified from Swartkrans. There are a few other fossils identified as *Homo* or 'early *Homo*' from Drimolen and Gondolin, most of which are only isolated teeth. A recent assessment by Darren Curnoe of the University of New South Wales, Australia, concluded that there are over sixty fossils attributed to early *Homo* in South Africa, constituting perhaps 40 per cent of the combined East and South African sample for these taxa (Curnoe 2006; see also Curnoe & Tobias 2006). However, the East African fossil sites have produced several fairly complete crania and two partial skeletons. Fossils attributed to early *Homo* are generally dated to less than 2.0 million years ago in South Africa, and like the robust australopithecines, they are thought to be evolutionary descendants from a gracile australopithecine species, but whether it was *A. africanus*, *A. afarensis* or some other species, and thus whether this evolutionary transition occurred in South Africa, East Africa or both (each proposal raises its own problems), has not yet been resolved.

The early members of the genus *Homo* can generally be distinguished from the australopithecines in that they possess larger brains, reduced tooth size (especially the premolar teeth), and have more orthognathic faces (the opposite of prognathic, being more vertical instead of projecting). Later members of the genus also demonstrate increased body size and further evolution of the locomotory adaptation for striding, modern human-like bipedalism. In addition, the first archaeological evidence for the manufacture and use of stone tools has been recovered from deposits associated with the earliest members of the genus *Homo* (though this may no longer be the case in East Africa). Not all of these features are found in all fossils attributed to the genus *Homo*, but they are generally considered to be the defining characteristics of the lineage.

South African palaeoanthropology has generally been concerned with these three groups of extinct hominids: gracile australopithecines, represented by *A. africanus*; robust australopithecines, represented by *P. robustus*; and early *Homo*, represented by both *H. habilis* and *H. ergaster*. Now we can consider each of these hominid groups from the Cradle in more detail.

Australopithecus africanus

The type specimen of *A. africanus* (that is, the first fossil so identified) was the 'Taung child'. Thus, the description of identifying characteristics for this species was originally based on a juvenile individual and it was imperative that adult specimens should eventually be brought to light. Robert Broom provided this evidence when he recovered the first adult *A. africanus* specimen from Sterkfontein in 1936. This specimen was catalogued as TM 1511 and consisted of a natural cranial endocast and a fairly damaged facial skeleton with most of the teeth. Additional fossils were soon recovered, including additional cranial material, mandibles and teeth. Broom determined that the Sterkfontein specimens were sufficiently different to warrant identifying a new australopithecine species, which he named *Plesianthropus transvaalensis*. Over the next couple of decades, over seventy additional specimens were recovered, including crania such as Sts 5 ('Mrs Ples') and Sts 71, and a partial skeleton referred to as Sts 14 which included a pelvis, vertebrae, ribs and the upper part of a femur.

James Kitching and his brothers recovered the first australopithecine specimen from Makapansgat in 1947; it consisted of the occipital region of a skull. Raymond Dart published the description of this specimen and gave it a new species name, *Australopithecus prometheus*. Later specimens such as the juvenile mandible MLD 2 and even the partial cranium MLD 37/38 were provisionally referred to this new taxon. The species name *prometheus* was a reference to the Greek hero who stole fire from the gods, and reflected Dart's earlier observation of the blackened bone in the Makapansgat deposits, leading him to postulate that the so-called 'troglodytes' possessed the capacity to produce and use fire as part of their cultural adaptation (much later, it was shown that the bones were blackened as a result of manganese staining from groundwater during fossilisation). These ideas were reflected in his elaborate osteodontokeratic (bone, tooth, horn) theory about the use of animal remains as cultural implements and weapons by australopithecines which he thought occupied the caves at Makapansgat.

In the 1960s JT Robinson revised the taxonomy for the australopithecines and delegated all the material from Taung, Sterkfontein and Makapansgat to the originally named species, *A. africanus*; this taxonomic arrangement is still in use today. In 1966, Phillip Tobias initiated renewed excavations at Sterkfontein, assisted by Alun R Hughes and then Ronald J Clarke. This ongoing excavation is still in operation, and many new and important fossils have been recovered and described. These include the partial skeleton Stw 431, the robust male cranium Stw 505 (referred to, of course, as 'Mr Ples'), and the recently discovered and nearly complete skeleton Stw 573, 'Little Foot', which has so far been assigned only to the genus *Australopithecus* – no species name for it has yet been determined. In all, the *A. africanus* assemblage from South African sites includes over 1 000 fossil specimens, demonstrating that *A. africanus* is one of the best-represented early hominid species anywhere in Africa – and is best represented at Sterkfontein.

Dart's original definition of *A. africanus* in 1925 identified a number of traits that he considered to be of evolutionary significance. At the time of the Taung discovery, it was thought that the earliest human ancestors would have a human-like brain (that is, a large cranial capacity), and ape-like teeth and postcranial structures. However, one of the remarkable aspects of the 'Taung child' is that it showed just the opposite pattern – a small brain well within the range of modern apes (approximately 400 cc as a juvenile) and features of the dentition, such as a reduced canine tooth crown, that were intermediate but reflected an evolutionary trend towards modern human morphology. When Dart succeeded in removing the mandible from the Taung cranium in 1929, it was also clear that the premolar and molar teeth were much more similar to humans than to apes such as the chimpanzee.

A. africanus is characterised by a relatively small brain (on average between 400–500 cc compared to roughly 400 cc in chimpanzees and 500 cc in gorillas; humans have an average cranial capacity of about 1 350 cc), and a prognathic or projecting face and jaws. Above the eye orbits they possess a bony ridge called a supraorbital torus, and a slightly rising forehead. However, the facial skeleton is still large compared to the braincase, and the overall appearance bears some resemblance to apes such as the chimpanzee.

In addition to the cranial remains described, *A. africanus* is represented by two partial skeletons from Sterkfontein, Sts 14 and Stw 431, which preserve bones of the upper and lower limbs, vertebral column and ribs, and pelvis. The new specimen Stw 573 is also informative about the locomotory behaviour of the genus, as are fossils of *A. afarensis* in East Africa. Postcranially, *Australopithecus* fossils demonstrate an intriguing mixture of traits suggesting that they were essentially bipedal from the waist down, but retained arboreal climbing anatomy of the upper limbs. In particular, the structure of the pelvis, hip joint and knee joint demonstrate all the essential features for bipedal, upright walking, though it is generally agreed that australopithecine 'bipedalism' was not exactly like that of modern humans – though the detailed nature of the differences is difficult to elaborate!

The complete structure of the foot is not well known, but Ron Clarke and Phillip Tobias have interpreted the 'Little Foot' skeleton from Sterkfontein (Stw 573) to demonstrate a great toe with some type of grasping capabilities, similar to apes and other primates. This also accords with Clarke's interpretation of the hominid footprint trail at Laetoli in Tanzania, which is attributed to the earlier taxon *A. afarensis*. Of course, not all researchers agree with this reconstruction – it will take the recovery of more complete foot fossils to further illuminate this issue.

While the lower limb structure of the australopithecines was essentially that of a biped, their upper limbs were relatively long, and they had elongated, curved hand and finger bones. These traits are all similar to those found in modern apes, and are useful adaptations for climbing trees and spending time moving around arboreally. It is not at all unusual for different primate species to combine arboreal climbing and terrestrial locomotion in pursuit of various food resources, but australopithecines were unique in that they used a bipedal mode of locomotion when moving through the terrestrial component of their habitat.

This is not to say, however, that any early hominid biped walked just like modern humans. Research by Henry McHenry and Lee Berger has elaborated on the significance of differences in upper- and lower-limb proportions in *A. afarensis* ('Lucy') and *A. africanus*. Based on comparisons of joint size (not limb length) in early hominids, modern humans and other primates, they concluded that *A. afarensis* had reconstructed limb joint size proportions more similar to modern humans, and the relative joint sizes for *A. africanus* are more chimpanzee-like – the joint sizes of the upper limb fossils from Sterkfontein are very large compared to the lower limb joints. While it is difficult to infer exactly what this means for locomotion (they were both still bipeds with 'chimp-like' limb-*length* proportions), these reconstructions also produce an unanticipated dilemma for reconstructing hominid phylogeny. *A. afarensis* lived earlier, and may have been ancestral to *A. africanus*, but had limb joint proportions similar to modern humans, while the joint sizes in *A. africanus*, which lived later and may have been descendant, were more similar to chimpanzees. This pattern of limb joint morphology does not

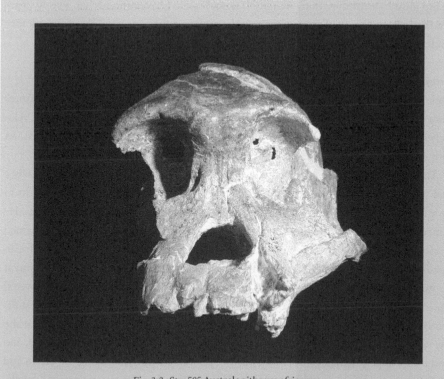

Fig. 3.3 Stw 505 Australopithecus africanus

reflect progressive or 'straight-line' evolution, unlike that of the brain, face and teeth in these hominids. This may mean that these two australopithecines are not related on the same lineage, or it may be a reflection of the mosaic and unpredictable nature of the evolutionary process.

Finally, it is important to realise that the australopithecines were much smaller than modern humans in overall body size. Reconstructed body mass for *A. africanus* generally falls between 30–40 kilograms, and reconstructed stature falls between 115–138 centimetres. A typical modern human adult weighs 55–65 kilograms, and could be 160–175 centimetres tall (these figures include variation between males and females). This difference in overall body size is a very important consideration when attempting to reconstruct locomotory function in early hominids, but it is difficult to clarify exactly what effect those differences would have had. For example, the hip and knee structure of *Australopithecus* is essentially bipedal, but the femoral head (at the hip joint) is relatively small, as observed in quadrupedal chimpanzees. Does this difference indicate something about locomotory function, or is it simply a reflection of the smaller body size of *Australopithecus* (because the hip joint transmits less body weight)? The fact is that early hominids had a body structure that was unlike any living primate, indicating that their exact mode of locomotory function was also unique – and thus difficult to describe in detail.

Paranthropus robustus

While the australopithecines as a group seemed to share some common general features when compared to the genus *Homo*, such as a small brain, larger cheek teeth, and relatively robust masticatory anatomy, the robust australopithecines were quite distinct from *Australopithecus africanus*.

The first fossil specimens attributed to robust australopithecines were discovered by a local schoolboy named Gert Terblanche at Kromdraai near Sterkfontein, and brought to the attention of Robert Broom, who visited the site and recovered additional fossils. He immediately noted that they were very different from the Sterkfontein fossils and named a new species, *Paranthropus robustus*. Assisted by the University of California's African expedition led by Wendell Phillips, Broom and Robinson were able to begin excavation at the nearby site of Swartkrans in 1948, and additional 'robust' material was recovered almost immediately. Broom felt that the Kromdraai and Swartkrans fossils were sufficiently different to warrant placing them in different species, and thus he named the new species *P. crassidens* for the Swartkrans fossils. This work continued until 1952, and produced a large assemblage of robust australopithecine fossils, including the SK 46 and SK 48 crania, the SK 12 mandible, and a variety of rather fragmentary postcranial fossils.

In 1959, Louis and Mary Leakey recovered the first 'hyper-robust' australopithecine cranium at Olduvai Gorge, Tanzania – catalogued as OH 5, and initially named *Zinjanthropus boisei*. This specimen and many later finds from East Africa demonstrated larger cheek teeth and even more robust masticatory structures – simply massive mandibles, extremely flared zygomatic (cheek) bones, and strong sagittal crests.

Beginning in 1965, CK Brain of the Transvaal Museum began what turned out to be approximately two decades of excavation and research at Swartkrans, producing additional fossil hominids and clarifying the stratigraphic and geological context of the cave deposits. Recently, starting in 2005, Brain and Travis Pickering have re-initiated excavations at this important locality.

Finally, excavation at a new site called Drimolen began in 1992, under the direction of Andre Keyser and Colin Menter. This extremely productive site has produced some ninety hominid fossils, mostly of *Paranthropus robustus*. This assemblage includes some very well-preserved specimens, including the first associated cranium and mandible (DNH 7), which is also recognised as the first female robust australopithecine from South Africa. Additional material includes a large (probably) male mandible (DNH 8) and numerous cranial, dental and postcranial fossils, including a number of juveniles.

The term 'robust' was originally coined because it was thought that such species had a considerably larger body size than their gracile 'cousins'. However, the most recent assessments of all the available *Australopithecus* and *Paranthropus* material indicate that differences in body size between the species were not so significant, but that sex dimorphism might have been quite pronounced in some species. On average, *Paranthropus* was probably only a few kilograms larger in body mass, and may have been no taller, than a typical *A. africanus* individual. However, it must be noted that the known sample of postcranial fossils attributed to *Paranthropus* is much smaller than that for *Australopithecus*, and there are no known partial skeletons for the genus.

The main differences between gracile and robust australopithecines are in the dental and masticatory structures of the cranium and jaws. The premolar and molar teeth of *Paranthropus* were, on average, much larger than those of *Australopithecus*. In addition, the supporting bony structures of the face and jaws were larger and heavily buttressed in order to sustain the heavy chewing forces generated during a *Paranthropus* meal. In particular, the robust morphology includes broadly flaring cheekbones (or zygomatic arches) and the presence (in males) of a sagittal crest atop the skull, to which the temporalis muscles attached. The temporalis muscle is the largest of the four muscles of mastication – the muscles that produce the biting force during chewing. These features gave *P. robustus* a broad, flat face that was relatively orthognathic in comparison to *A. africanus*. In combination with this assessment of australopithecine anatomy, studies of dental microwear have suggested that

Paranthropus was dentally specialised for a diet that required heavy chewing, or mastication, probably indicating that it consisted of hard and relatively low-quality foodstuffs, such as underground tubers, roots or coarse browsing vegetation. Another possibility is that it consumed pulpy fruits with a fibrous outer skin.

Thus, the conclusion from anatomical comparisons and other evidence is that *Australopithecus* and *Paranthropus* occupied different dietary and adaptive niches. However, the detailed nature of these differences is elusive, and recent studies (see below) indicate that there may have been considerably more overlap in diet and habitat than previous interpretations would have allowed. Future research will certainly be focused on resolving these issues.

Homo habilis

The species *Homo habilis* was first named and described in 1964 by Louis Leakey, Phillip Tobias and John Napier, based on several fossil specimens from Olduvai Gorge thought to be dated to about 1.75 million years ago, and contemporaneous with the *Zinjanthropus* (*Paranthropus*) skull already described. In the original description, this species was distinguished from *Australopithecus* in having a slightly larger brain (approximately 600 cc or larger), smaller cheek teeth, and a more human-like hand and other differences of the postcranial skeleton associated with bipedalism. At the time, no South African fossils representing *H. habilis* were known.

The earliest evidence for members of the genus *Homo* in South Africa is found at Sterkfontein in deposits currently interpreted by Ron Clarke and Kathy Kuman to be intermediate between Member 4 (*Australopithecus*-bearing) and Member 5 (tool-bearing) deposits, probably dating between 2.0 and 2.4 million years ago. Traditional ideas are that the species *Homo habilis* is best represented by the partial cranium Stw 53, recovered in 1976 by Alun Hughes. Many additional fossils from sites in East Africa have also been attributed to this taxon, in addition to those originally described from Olduvai Gorge. These include fossils from Koobi Fora, Kenya, which demonstrate a high degree of variability. KNM ER 1470 is a large-brained specimen (775 cc) with a flat and broad facial skeleton, while KNM ER 1813 is small-brained (510 cc) and has other australo-pithecine-like affinities.

Thus, *Homo habilis* is an important yet problematic taxon for a number of reasons. Firstly, this species is recognised as the earliest member of the genus *Homo*, and it demonstrates a suite of characteristics that distinguish it from the gracile australopithecines from which it is thought to have evolved. However, *H. habilis* is by no means thought to be 'human'. For example, it possesses a larger brain and reduced teeth compared to any australopithecine, indicating that the species had evolved to occupy a very different adaptive niche compared to its ancestors. However, its brain was probably no larger than 750 cc – just over half that of a modern human – suggesting that human-like higher brain functions had yet to evolve. In addition, based primarily on the OH 62 partial skeleton recovered from Olduvai Gorge in 1986 by Don Johanson and his colleagues, *H. habilis* appears to have retained the relatively small body size and relatively long arms of species such as *A. africanus*. It was bipedal, but still retained some arboreal capabilities.

Secondly, the group of fossils referred to is extremely variable, and some researchers have divided the *H. habilis* hypodigm (the total assemblage of fossils attributed to this species) into two taxa, recognising *H. rudolfensis* as a second species existing at about the same time in East Africa. *H. rudolfensis* differs from *H. habilis* in having a larger brain (usually over 700 cc) but larger, more australopithecine-sized chewing teeth, as typified by the KNM ER 1470 cranium. In this stricter sense, then, *H. habilis* typically had a smaller brain, but also smaller dentition and a less robust cranium, as typified by the KNM ER 1813 cranium and others. Given these mosaic combinations of cranial and dental features, which species is most likely to have been the direct ancestor to later species of *Homo*, and ultimately to our own *H. sapiens*? This issue is far from resolved, and will certainly stimulate much interesting future research.

The fossil record for *H. habilis* in South Africa is very scanty, and no fossils have been attributed to the larger-brained *H. rudolfensis* (though some specimens from the Malawi Rift in southern Africa are thought to represent this taxon). This is significant in that one proposed phylogenetic model for early hominid evolutionary relationships places the South African taxon *A. africanus* as the ancestral species to the genus *Homo*. Temporally, this certainly is plausible, since the known time ranges of *A. africanus* and *H. habilis* overlap between approximately 2.4 and 2.0 million years ago – ancestors and their descendants certainly should live in the same geographic region, and at similar time periods! In contrast, the alternative model places the East African taxon *A. afarensis* as the ancestor to *Homo*, but this leaves a temporal gap from approximately 3.0 million years ago to approximately 2.4 million years ago for which some plausible account must be made. A recent find in Ethiopia by Berhane Asfaw, Tim White and their colleagues, called *Australopithecus garhi*, is thought by some to fill this gap, and it is dated to about 2.5 million years ago.

However, Ron Clarke considers the Stw 53 cranium to be a 'derived australopithecine' rather than a true member of the genus *Homo*. If this model were validated, it would indicate that early *Homo* had not yet evolved at the time that Member 5 at Sterkfontein was deposited, but it also suggests that australopithecines were manufacturing and using stone tools before 2.0 million years ago – and thus that tool-using adaptations should not be considered to be a hallmark of the genus *Homo*. A recent reassessment and new reconstruction of Stw 53 by Darren Curnoe and Phillip Tobias supports the original designation to *H. habilis*, suggesting that this taxon existed in both East and South Africa. Either way, the Stw 53 cranium provides important evidence for a proposed *Australopithecus–Homo* transition in South Africa. As is so often the case in palaeoanthropology, additional fossil specimens are needed to fill in the gaps and help resolve these issues.

Homo ergaster

In 1949 Robert Broom recovered a nearly complete mandible from Swartkrans designated SK 15, which he placed in a new genus and species, *Telanthropus capensis*. Compared to the australopithecines he had recovered from Swartkrans and other sites, this mandible and the teeth it contained were more lightly built and much smaller in size.

The species *Telanthropus capensis* is now called *Homo ergaster*, which is also represented by a partial cranium (SK 847) from Swartkrans. This taxon is generally considered to be an African form of the Asian species *H. erectus*, represented by the material recovered in Java in the 1890s and in China in the 1920s and later. However, the relationship between *H. ergaster* and earlier forms of *Homo* in Africa, or Asian populations of *H. erectus*, is extremely controversial. Some researchers would, in fact, simply call all of it *H. erectus* (the original species name) and explain the morphological differences observed as regional variation – in other words, local adaptations to environmental variation.

In East Africa, the taxon *H. ergaster* is represented by a number of well-preserved crania, and a nearly complete cranium, particularly at Koobi Fora in Kenya. These fossils are dated between 1.8 and 1.5 million years ago, and are also contemporaneous with some of the later *Paranthropus* fossils from sites such as Koobi Fora, Omo and Olduvai Gorge. As already mentioned, it is generally thought that *Paranthropus* had occupied a more specialised, but perhaps marginal habitat, while *H. ergaster* is thought to have been much more of a generalist, occupying a variety of habitats and subsisting on a diverse diet that included a substantial amount of animal protein. Based on interpretations of the stone tool assemblages, and on cut-marked animal bone found at many sites, *H. ergaster* is thought by some researchers to have been the first truly hunting hominid, thus exploiting a dietary niche that no other hominid had occupied.

H. ergaster morphology was very different from earlier hominids, including its probable ancestor *H. habilis* (here including *H. rudolfensis*). It had a much larger brain, approaching 1 000 cc, and its teeth were essentially modern though larger in size. Its facial skeleton was reduced compared to *Australopithecus*, but retained a heavy brow ridge and slight prognathism. Perhaps the most striking difference, however, was in its body size and limb proportions.

The partial skeleton KNM WT 15000 from Nariokotome, Kenya, in East Africa provides some important clues regarding the adaptations of this taxon. Though the individual represented by this skeleton was still a juvenile, probably no more than 13 years old, it demonstrates that *H. ergaster* had evolved modern human limb proportions – longer legs, shorter arms, and a tall, lean physique. It appears to have been bipedal in the modern sense, and the arboreal climbing adaptations observed even in *H. habilis* (OH 62) were lost. KNM WT 15000 was approximately 1.68 metres tall and probably weighed about 60 kilograms – and would have been taller and heavier as an adult. By this time, *H. ergaster* had evolved an essentially modern body.

It is thought that these features of the *H. ergaster* body are an adaptation for a warmer, more arid climate that had evolved after 2.0 million years ago, and for the more active terrestrial lifestyle that would be required for a hunter. The evolution of this essentially modern body in *H. ergaster* is seen as a sharp departure from earlier hominid species, including *H. habilis*. This is one factor that prevents clarification of evolutionary relationships during this important time period.

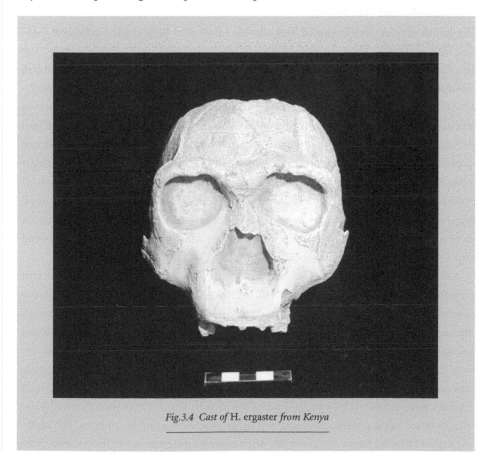

Fig.3.4 Cast of H. ergaster *from Kenya*

H. ergaster is typically described as the first hominid to migrate into Asia – an event sometimes referred to as 'Out of Africa I'. However, there are certain problems with this scenario relating to morphological variation, associated stone tools, and the fact that some of the earliest Asian sites are essentially as old as those in East Africa. In addition, one of the earliest sites 'Out of Africa' – that at Dmanisi in the Republic of Georgia – has produced several fossil skulls that are more primitive and smaller-brained than *H. ergaster* in Africa, and have been compared by some to *H. habilis*. For these and other reasons, some researchers, such as Robin Dennell (2003; Dennell & Roebroeks 2005), have postulated that *H. ergaster* actually represents a population that left Africa earlier, evolved in an Asian context, and then re-entered Africa at about 1.8 million years ago in a subsequent (and reverse) migration. As usual, much further work, and more new fossils, will be required to resolve such questions.

Controversies and conclusions

The South African early hominid sites and the fossils they have yielded provide significant information for the scientists working to resolve the many important questions that arise in palaeoanthropology. For many years, the bulk of evidence about hominid evolution in South Africa has come from only two sites – Sterkfontein and Swartkrans – where long-term excavations were being conducted. However, more recent work at numerous fossil sites, both within the area designated the 'Cradle of Humankind' and elsewhere, has demonstrated the potential for new fossil finds and for productive research in South African palaeoanthropology. It is also important to remember that palaeoanthropology is not just about hominid fossils themselves; a great deal of contextual information is required in order to properly analyse such intriguing fossil specimens. In this sense, research at a variety of South African fossil sites, using an increasing range of new analytical techniques, has contributed to advances in several important areas of palaeoanthropological research.

One of the great difficulties scientists have faced historically in interpreting South African Plio-Pleistocene hominid sites has been the lack of a reliable source for absolute chronological dates for the dolomite cave deposits. While in East Africa the numerous volcanic cave tuffs provide radioactive isotopes for techniques such as potassium-argon dating, no such deposits are present in South Africa. Until recently, faunal correlations with dated East African site deposits were the most widely used technique for dating sites such as Sterkfontein and Swartkrans. In other words, dates for South African deposits were estimated by comparison of fauna with East African sites that had been reliably dated using potassium-argon and other geological aging techniques. The problem is that different fauna at a given site often produce different age estimates, and there is no guarantee that different species necessarily lived in both East and South Africa at exactly the same time period.

Another commonly used method is palaeomagnetic dating, which relies on reference to the Global Polarity Timescale, and is also a relative dating technique involving some uncertainty in the age estimates produced. This problem has long been a hindrance to obtaining reliable interpretations of South African sites, since the geological age of a site or of a region is a fundamental aspect of the framework needed for understanding evolutionary relationships. It has thus long been a desire of researchers to develop an absolute dating technique for use on South African deposits.

The associated skeleton and skull from Member 2 at Sterkfontein, Stw 573 or 'Little Foot', discovered in 1994, can be viewed as something of a case study in this regard. It was initially dated using palaeomagnetic dating methods to approximately 3.5 million years ago; a new technique called cosmogenic dating later produced an age estimate of at least 4.0 million years ago. Additional new material from a Sterkfontein locality called Jacovic Cavern has also been assigned a comparably early date. The dates of these deposits are of great importance because while Ron Clarke has identified the Stw 573 skeleton as an australopithecine, he has not attributed it to *A. africanus* because he has observed some primitive morphological features. The earlier date is thus consistent with this provisional taxonomic assessment. Recent controversy over the cosmogenic dates has arisen from age estimates produced using other techniques, which suggest that the deposits are approximately 1 million years younger – that is, within the conventional age range for *A. africanus*. While morphology is ultimately the basis for taxonomic assignments, correctly resolving the date will play a role in debates over this specimen's taxonomic and phylogenetic placement.

A second issue involving recent controversy is that of early hominid dietary reconstructions, especially the conventional interpretations for distinct dietary niches between *Australopithecus* and *Paranthropus*. This issue is extremely important because it is the basis for interpretations of morphological and adaptive differences and related aspects of life-history variation in these genera.

Dietary assessments for extinct species are conventionally made using a combination of information about the species' morphology, dental microwear, and site palaeoenvironment. However, more recent studies of stable carbon isotopes also provide some insight into the dietary adaptations of these early hominids. When animals consume plants, they incorporate different carbon isotopes in the same proportions as those produced by the plant's photosynthetic pathway – the way in which the plant utilises carbon dioxide and water to produce sugar and oxygen for energy. The two most common such pathways – or methods of energy conversion – are known as 'C_3' and 'C_4', and the plants are thus known as C_3 and C_4 plants. The former include trees, shrubs and bushy vegetation occupying more closed environments, and the latter primarily include grasses and sedges that occur in more open and arid environments. Thus, animals feeding on either plants or other animals occupying (and feeding in) such vegetation zones will incorporate the same stable carbon isotopes into their hard tissues (bones and teeth), the proportions of which can still be measured in fossilised materials. In this way, it is possible to determine whether an extinct taxon, such as *Australopithecus* or *Paranthropus*, consumed plants located in (or other animals or insects feeding on) C_3 or C_4 vegetation communities.

What is interesting about these studies is that isotopic signals obtained from *Australopithecus*, *Paranthropus*, fossil baboons and even early *Homo* suggest that they all relied significantly on C_4 plant foods obtained in an open habitat. This indicates that early hominids and a variety of extinct primates were occupying similar feeding habitats in the Plio-Pleistocene, and must thus have been utilising their environmental resources in different and complex ways in order to avoid direct competition. Drawing on our understanding of modern primate diets, we can say that this may have involved eating different plants in the same habitat, seasonal differences in diet, and/or eating different proportions of the same foods. Further studies are clearly needed in order to clarify early hominid dietary differences.

A final issue demonstrates how the use of new technology can stimulate novel interpretations of existing fossil specimens, and involves the famous Sts 5 cranium, 'Mrs Ples'. Some researchers have observed that this skull presents unexpected morphology for a female *Australopithecus*. For example, it has very prominent anterior pillars, which are generally associated with having long canine roots, and presumably large canine crowns – a male trait in a sexually dimorphic species. Francis Thackeray of the Northern Flagship Institution, Transvaal Museum, has conducted some intriguing research using non-destructive imaging techniques such as high-resolution CT scanning of the Sts 5 cranium and its associated rock matrix (breccia). In one study, Thackeray observed that the rock matrix encasing the Sts 5 skullcap showed a calcite ridge aligned with the sagittal suture, thus resembling the sagittal crest observed in male apes for attachment of enlarged temporalis muscles used for chewing. While Thackeray could not conclusively determine the nature of this calcite feature in the Sts 5 breccia, some male australopithecine crania do demonstrate the presence of sagittal cresting. Thus, if such a structure were actually present on the Sts 5 skull, it would indicate that the individual were male – it is not present in females of any primate species.

In a subsequent study, Thackeray and his colleagues obtained high-resolution CT scans in the region of the tooth roots of Sts 5, and found that the foramen at the root apex of the permanent maxillary canines and M3 may not yet have been completely formed. This would indicate that Sts 5 represented a sub-adult individual, meaning that any sex-specific traits such as greater robusticity, including the presence of a prominent sagittal crest, would not yet have fully developed.

While this case does not provide firm and conclusive answers to some of the questions posed, it is an excellent example of the process of scientific enquiry. New techniques allow researchers to record observations that previously were impossible, and to grapple with questions that previously could not even be addressed. Every new line of evidence adds in some way to our knowledge of the fossil specimens, and in time confirmation of one or the other option will be possible. In addition, this kind of research allows us to incrementally put the flesh back onto the bones of these fossil ancestors – rather than just another old skull, we can now envision Sts 5 as 'Mrs Ples', 'Mr Ples', or perhaps 'Master Ples', the young australopithecine. Each interpretation carries different implications about the individual represented by the fossil skull.

In conclusion, the fossils and sites from the 'Cradle of Humankind', and from related areas, are important for a number of reasons.

Firstly, these sites have produced an abundance of fossils – over 1 000 specimens at current count – especially of *A. africanus* and *P. robustus*. In fact, these fossil hominid taxa are only known from a handful of sites in the Cradle and related areas.

Secondly, sites such as Sterkfontein and Swartkrans consist of long-term deposits, representing time spans of hundreds of thousands or even millions of years. Thus they have the potential to document evolutionary trends and transitions in the context of a regional environment sampled by a single cave site.

Thirdly, the sites of the Cradle document evidence for the manufacture and use of both bone and stone tools, in addition to earlier deposits that appear to represent 'pre-tool' periods. Chapters 6 and 7 describe the tool finds, and the evidence they provide about evolving ways of life in the region, in greater detail.

Finally, multidisciplinary research at these sites has the demonstrated potential to provide a variety of contextual information about hominid evolution – fossils of associated fauna, pollen and plant fossils, stable isotopes, and important chronological estimates that improve our ability to produce reliable and testable hypotheses about early hominid adaptation and evolution.

In this way, the fossil sites from the Cradle allow researchers to address a variety of questions about evolution, adaptation, variation and phylogenetic relationships of early hominids. The complete story will someday come from a comprehensive assessment of sites and fossils from both South and East Africa – and wherever else in the world new fossil discoveries are made.

SELECTED SOUTH AFRICAN EARLY HOMINID FOSSILS FROM SITES IN THE 'CRADLE OF HUMANKIND'

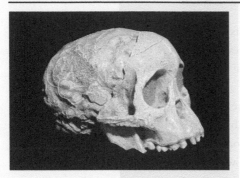

Description

Holotype of *Australopithecus africanus*. Well-preserved cranium of a juvenile, including a complete facial skeleton, partial mandible, dentition, and partial cranial vault with a natural endocranial cast.

Date discovered

1924

Museum Number	Taung 1
Site	Taung

Museum Number	TM 1511
Site	Kromdraai

Description

Holotype of *Paranthropus robustus*. Adult partial cranium with associated fragmentary mandible, including permanent teeth; some postcranial fragments associated.

Date discovered

1938

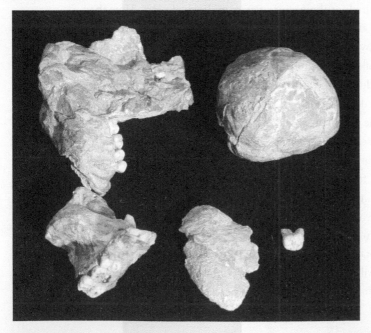

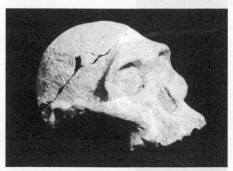

Description
Nearly complete adult cranium of
A. africanus locking the maxillary
dentition.

Date discovered
1947

Museum Number	Sts 5
Site	Sterkfontein

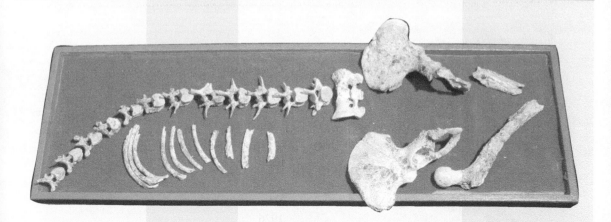

Museum Number	Sts 14
Site	Sterkfontein

Description
Partial skeleton of *A. africanus*,
including a partial left femur, nearly
complete pelvis, many vertebrae and
a few rib fragments.

Date discovered
1947

Museum Number	Sts 71
Site	Sterkfontein

Description
Partial cranium of *A. africanus*;
consists largely of the right side of
the skull.

Date discovered
1947

Museum Number	Stw 53
Site	Sterkfontein

Description
Partial cranium attributed
to *H. habilis*; includes
fragmentary cranial vault,
maxillae, heavily worn
teeth.

Date discovered
1976

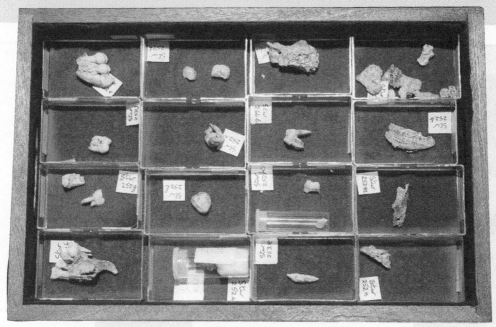

Museum Number	Stw 252
Site	Sterkfontein

Date discovered
1984

Description
Partial cranium of *A. africanus* or a related second species. Very fragmentary cranium, but including complete maxillary dentition.

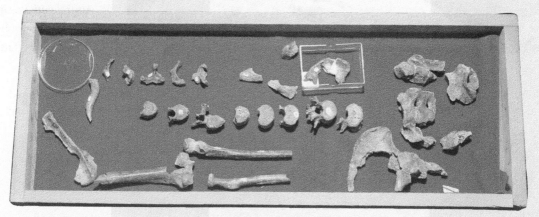

Museum Number	Stw 431
Site	Sterkfontein

Date discovered
1987

Description
Partial skeleton of *A. africanus*, including upper and lower limb bone fragments, pelvis, vertebrae, scapula.

Description
Partial but relatively complete cranium of a probable male *A. africanus*. Damaged and distorted, but including most of the cranial vault and facial skeleton.

Date discovered
1989

Museum Number	Stw 505
Site	Sterkfontein

Museum Number	Stw 573
Site	Sterkfontein

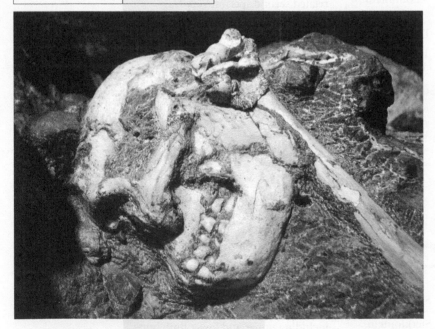

Description
A virtually complete skeleton (cranium, mandible, and postcrania) of an *Australopithecus* individual, commonly referred to as 'Little Foot'.

Date discovered
1980; 1994; 1997

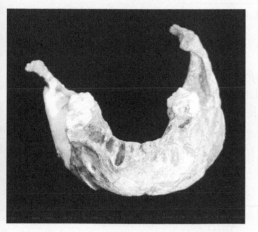

Museum Number	SK 15
Site	Swartkrans

Description
A fairly complete mandible of
H. ergaster (formerly referred to as
Telanthropus capensis).

Date discovered
1949

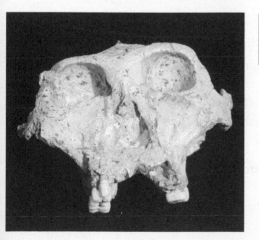

Museum Number	SK 48
Site	Swartkrans

Description
A fairly complete but partly crushed
adult cranium of *Paranthropus robustus*, including most of the dentition.

Date discovered
1950

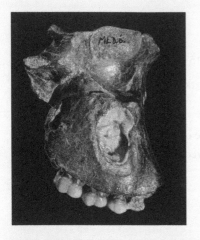

Museum Number	MLD 6
Site	Makapansgat

Description

A facial fragment of *A. africanus* consisting of a nearly complete right maxilla with several teeth.

Date discovered

1948

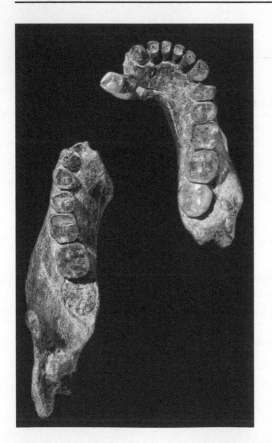

Museum Number	MLD 18 / MLD 40
Site	Makapansgat

Description

A left and right mandibular corpus and complete symphysis, including well-preserved teeth of adult *A. africanus*.

Date discovered

1953

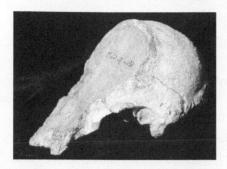

Museum Number	MLD 37/38
Site	Makapansgat

Description
Specimen comprising two joined pieces; nearly complete *cranial* vault of *A. africanus* with virtually the whole face missing.

Date discovered
1958

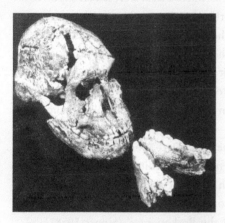

Museum Number	DNH 7 & DNH 8
Site	Drimolen

Description
A complete adult cranium and mandible of probable female *P. robustus*. A mandible of an adult probable male *P. robustus*.

Date discovered
1994

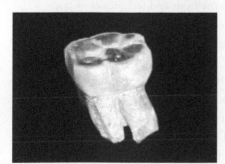

Museum Number	GDA 2
Site	Gondolin

Description
Left mandibular second molar of *Paranthropus* species.

Date discovered
1997

*NB. There are over 600 individual specimens from the Sterkfontein excavations alone; this table represents just a few of the well-preserved and important finds from the major South African Plio-Pleistocene fossil hominid sites.

UNRAVELLING THE HISTORY OF MODERN HUMANS IN SOUTHERN AFRICA: THE CONTRIBUTION OF GENETIC STUDIES

Himla Soodyall and Trefor Jenkins

Anthropological genetics in South Africa

Raymond Dart, in his writings as well as in the influence he exerted on his students, exhibited a deep interest in the earliest history of the living peoples of southern Africa. Using the methods of his time, Dart searched for pure strains, physical types and racial features among various populations. Accompanied by distinguished scholars like LF Maingard, P Kirby, CM Doke, ID MacCrone and Dorothea Bleek, Dart led an expedition to the Kalahari in 1936, spending a month working among the ?Auni≠Khomani San in the northern Cape province (now the Northern Cape), where it borders on Botswana and Namibia. Face masks were made and photographs were taken of seventy-seven individuals and are still housed at the Wits Medical School, where they are referred to as the Dart Collection. In addition, genealogical information was collected from each individual, every conceivable measurement was taken and their non-metrical cephalic features were recorded.

Adopting the typological approach popular at the time, Dart (1937a, b) advanced the opinion that the San exhibited Mongoloid, Armenoid and Mediterranean features, in addition to the Bush and 'Boskopoid' features which, in his view, predominated in the 'Bushmen and Hottentot peoples of today'. It was some years before the Cape Town anthropologist Ronald Singer exploded the myth of the Boskop 'race' (Singer 1958). The typological approach has been superseded by the methods and techniques of population genetics.

When the Viennese scientist Karl Landsteiner demonstrated in 1900 that all human beings could be assigned to one of a small number of blood groups by a simple test on their red blood cells, he opened up a field of research which continued and expanded throughout the twentieth century. The genetic diversity exhibited by the peoples of southern Africa became an interest of South African medical practitioners as early as 1921, undoubtedly because of its relevance to blood transfusion practice. The studies came within a couple of years of the demonstration in 1919 by Ludwig Hirschfeld and his wife, Hannah, that the frequencies of the genes determining the ABO blood groups varied between populations (Hirschfeld & Hirschfeld 1919). The Hirschfelds served as medical officers during the First World War in Salonika in north-eastern Greece, where they provided medical care for Serbs and the Allied expeditionary forces retreating from their defeat at Gallipoli. Here they had access to many different ethnicities – not only French, British and other soldiers from Europe, but also soldiers from French and British colonies in Asia and Africa (Senegalese, east Indians, Malagasy, Jews and Arabs). The results of the 'English' soldiers tested, when broken down, revealed that this group included individuals of whom twenty-nine were Welsh, fifty-two Scots and sixteen Irish in origin! Their pioneering contribution to the field demonstrated that the three common genes of the ABO system, gene *A*, gene *B* and gene *O*, were found in *all* populations, but the *frequencies* of these genes varied from one population to another. For example, whereas gene *A* has a higher frequency than gene *B* in all the European populations, gene *B* has a higher frequency than gene *A* among the Senegalese and among Indians.

The Hirschfelds believed that they had shown 'that experiments in immunisation deserve to be made use of for the solving of anthropological questions', and went on to affirm that 'a close co-operation would be necessary between anthropologists and serologists and the researches should be conducted on an international basis' (Hirschfeld & Hirschfeld 1919: 679). Mourant acknowledged the importance of this study 'not simply as making the discovery of one particular anthropological character but as being the first application to anthropology of a totally new method, the study of gene distributions: since there was no necessary distinction between the individuals of one population and of another, the populations themselves became the units of study' (Mourant 1961: 155).

Within fifteen months of the Hirschfeld's report, Harvey Pirie, a microbiologist who had joined the staff of the South African Institute for Medical Research (SAIMR) in 1918, reported on the frequencies of the ABO blood groups in 250 South African Bantu speakers (Pirie 1921). Adrianus Pijper (1929–30, 1932, 1935) extended these studies to include data on the Bantu speakers and the San and Khoi from Namibia (formerly South West Africa). Elsdon-Dew, working at the SAIMR in the 1930s, took a special interest in blood groups as they applied to forensic medicine and presented results on blood groups from thousands of individuals from southern Africa, Central Africa and East Africa in two monographs (Elsdon-Dew 1936, 1939), after travelling many thousands of miles across southern and Central Africa to collect the samples and to test them in the field.

As more blood group systems like Rhesus, MNS, Kell and Duffy were discovered in the 1940s and 1950s, Alwyn Zoutendyk, working at the SAIMR, and Maurice Shapiro from the South African Blood Transfusion Service, incorporated these systems in extending population studies among the Bantu-speaking, Khoikhoi and San groups. Shapiro (1951) claimed that the various Bantu-speaking 'tribes' or chiefdoms constituted a homogenous group, and Zoutendyk and co-workers (1953, 1955) showed quite convincingly that the San and Khoikhoi were African and not influenced by Asians as had been suggested by Dart (1951).

Dart did not ignore the 'new anthropology', and presented a paper entitled 'African Serological Patterns and Human Migrations' as his presidential address to the South African Archaeological Society in Cape Town in March 1951. Dart described the exercise as 'a quest, a search for the real past of this vast continent'. His interpretation of the global distribution of the ABO blood groups was naïve, not taking into account the fact that the gene frequencies in particular populations may have been determined by natural selection as well as by migrations.

The dawn of the molecular genetics era in South Africa

Peter Brain, a South African working as a medical officer at Shabani mine in southern Rhodesia (now Zimbabwe) in the mid-1950s, had an interest in sickle-cell anaemia which led him to hypothesise that possession of the gene in single dose conferred some protection against malaria (Brain 1952a, b, 1953, 1956). Trefor Jenkins, who had worked as a mine medical officer at Wankie (now Kwange) in Zimbabwe, was intrigued by sickle-

SAMPLE COLLECTION AND GENETIC STUDIES ON THE SAN POPULATION FROM THE KALAHARI

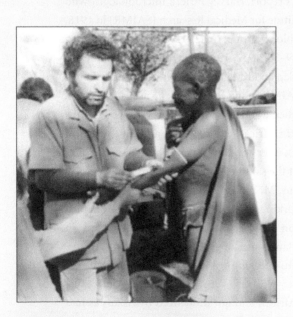

Trefor Jenkins has conducted research on the San ('Bushmen') since 1963. This has involved collecting blood samples from volunteers living in the Kalahari. Over the years these blood samples have been used to conduct various types of genetic research, first using the classical markers that were popular in the 1960s and into the 1970s, and then using DNA markers which became available in the 1980s.

The immunoglobulin haplotype $Gm^{1,13,17}$ was found to be a good Khoisan

Fig. 4.1 Trefor Jenkins collecting a blood sample from a San individual in the Kalahari

cell anaemia and its clinical manifestations in African children – he showed that it differed from that form of the disease seen in African-Americans (Jenkins 1965; Jenkins et
al. 1968). Linus Pauling, in 1949, had used electrophoresis to separate sickle-cell haemoglobins from normal haemoglobins, thereby discovering the biochemical basis for sickle-
cell anaemia and labelling it a 'molecular disease' – the first of many to be identified
in subsequent years. Having noted the high prevalence of sickle-cell anaemia among
children of immigrant Malawian workers at Wankie, Jenkins reported a series of cases
(Jenkins 1963) and went on to report the frequencies of the sickle-cell trait in various
populations, including the Gwemba Valley Tonga who, with other Zimbabwean populations, had very low frequencies of the trait.

Jenkins moved to South Africa in 1961 to work as a learner surgeon in Durban.
While there, he heard from a colleague, a University of the Witwatersrand graduate, of
an 'eminent teacher of anatomy at Wits' by the name of Phillip Tobias, and was even
more intrigued to hear that Tobias had conducted fieldwork among the Tonga people of

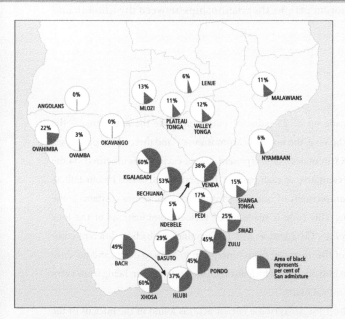

*Fig. 4.2 Map of southern Africa showing the estimated percentage
of San admixture in Bantu-speaking groups as determined
by the frequency of $Gm^{1,13,17}$ (From Jenkins et al. 1970).*

genetic marker, and was used to estimate
the proportion of Khoisan gene admixture in Bantu-speaking groups in southern Africa (Jenkins et al. 1970). It was
estimated that more than 50 per cent
of genes in the Tswana and just under
50 per cent in Cape Nguni groups were
derived from the Khoisan as a result of
gene flow from them. The other Bantu-
speaking groups of South Africa have
lower but still appreciable proportions
of these genes, whilst the proportion is
significantly lower in Bantu-speaking
populations in Namibia, Zambia and
Mozambique. The Angolan Bantu
speakers and the Kavango in Namibia
did not have this haplotype at all.

the Gwembe (Zambezi) Valley on the northern shore of Lake Kariba. Jenkins visited Tobias and was offered a demonstrator position in the Anatomy Department, which he took up in February 1963.

The advancement of the 'new anthropology' in South Africa from the early 1960s developed into the modern discipline of molecular anthropology, spearheaded by Phillip Tobias and by Stan Blecher and Trefor Jenkins, who carried out their initial field-work among the San in 1963 (Jenkins et al. 1968). During 1964 and for many years thereafter, Jenkins carried out population genetics research among the San, and subsequently he and his colleagues, in particular George Nurse at the SAIMR, continued to make use of a wide range of classical gene markers – blood groups, serum proteins and red-cell enzymes, as well as lactose tolerance, colour blindness and PTC tasting profiles – to examine the genetic affinities of the peoples of southern Africa (South Africa, Namibia, Botswana, Angola, Zambia) and, when it became possible, extended these studies to populations from neighbouring countries further to the north (Nurse et al. 1985). These genetic methods made it possible to examine questions on the relationship between the San (formerly 'Bushmen') and Khoikhoi (formerly 'Hottentots'), often referred to collectively as the Khoisan people; the relationships between the Khoisan and

INHERITANCE

Humans have 46 chromosomes made up of 23 pairs; 22 pairs constitute the autosomes – the chromosomes excluding the X and the Y, i.e. the non-sex chromosomes – and there is one pair of sex chromosomes (XX in females and XY in males). One representative of each pair is inherited via the egg from our mothers and one representative of each pair via the sperm contributed by our fathers at fertilisation. The total genetic complement of one set of human chromosomes contains some 3 billion letters of the genetic code. This code is made up of various combinations of the four nucleotides – A (Adenine), C (Cytosine), G (Guanine) and T (Thymine) – which in different combinations control the development of the organism from conception to birth to death, and produce the genetic variation that confers the individuality which distinguishes each human being from every other one, with the exception of identical twins.

Chromosomal DNA (DNA is short for deoxyribose nucleic acid) is found in the nucleus of the cell and is referred to as nuclear DNA. But there is a small amount of DNA which comes to us in the mitochondria (small organelles in the cytoplasm of our cells which are essential for the energy production for that cell). Mitochondria, including their DNA molecules, exist in every cell of the body,

Bantu-speaking people; and the genetic relationships of the different chiefdoms of Bantu speakers. They also permitted elucidation of the genetic constitution of different, so-called 'coloured' populations of South Africa; and assessment of the extent of gene flow between the different populations of the sub-continent.

Molecular evolutionary genetics and African origins

The discovery of the double helical structure of DNA by James Watson and Francis Crick in 1953 heralded the birth of a new discipline, molecular biology. Another twenty-five years were to elapse before the discovery of the first human DNA polymorphism. 'Polymorphism' is the term used by geneticists to refer to the occurrence in a population of different forms of a gene. These changes or mutations occur purely by chance. Some of these may provide a selective advantage on the individual, in which case they will be retained and may increase in frequency. Further developments in the field, including DNA sequencing and the ability to study variation, using very small amounts of DNA extracted from a blood sample or a cheek swab, revolutionised the field of molecular genetics and physical anthropology.

Fig. 4.3 *Schematic diagram of a cell showing nuclear DNA in the nucleus and mitochondrial DNA in the mitochondria*

but they are passed on to the offspring only in the cytoplasm of the ovum, which naturally comes only from the mother. This mitochondrial DNA (mtDNA) can be studied in detail in the laboratory after being extracted from a blood sample, from a cheek swab or from a host of other sources (tooth, bone, blood stain, semen, hair, etc.). Because of its unique mode of inheritance through the maternal line only, there cannot be recombination because it does not exist in a paired form. When it passes from mother to child it only rarely changes its DNA sequence, and therefore provides a unique insight into our ancestry through the matrilineal line.

MITOCHONDRIAL DNA AND HUMAN ORIGINS

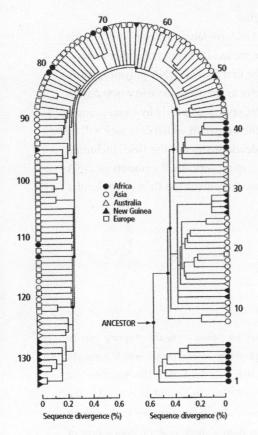

Fig. 4.4 *Phylogenetic tree showing the relationship of mitochondrial DNA lineages found in individuals sampled from Africa, Asia, Australia, New Guinea and Europe. The deepest branch in the tree that connects to the hypothetical ancestor contains mitochondrial DNA found exclusively in Africans, with lineages from people from the rest of the world being placed on shallower branches on the tree; this has been used as evidence to support the 'Out of Africa' theory concerning modern human origins.*

Diversity within mtDNA was initially studied by restriction fragment length polymorphism (RFLP) mapping. This was done by means of the extraction of mtDNA, in the original study, from placentae, and then analysing the DNA to define the variation existing among the individuals. Researchers working with Alan Wilson at Berkeley in the USA first adopted this approach. Using about eighteen different restriction enzymes, Cann et al. (1987) were able to examine about 20 per cent of the 16 596 base pairs or building blocks of DNA that make up the mitochondrial genome.

Cann and colleagues constructed a phylogenetic tree linking the lineages found in the sample of 147 individuals they examined, and dated the tree using the mtDNA mutation rate. The tree suggested a root from a single woman who lived in Africa around 200 000 years ago.

The descent of all mtDNAs from one woman very quickly stimulated the 'mitochondrial Eve' controversy. Criticisms focused on the use of African-Americans (only two of the twenty individuals used were born in sub-Saharan Africa) as representatives of African mtDNAs; the method used to generate the tree, which was questioned since it did not guarantee the most parsimonious tree; and the method of midpoint rooting used to locate the root, which placed it at the midpoint of the longest branch – outgroup rooting was the preferred method. Despite these criticisms, subsequent studies, including data on whole mtDNA sequences in over 500 individuals, have supported the major conclusions of Cann et al. (1987), i.e., that humans have a recent African origin for mtDNAs.

In 1987 Rebecca Cann, Mark Stoneking and Alan Wilson published a seminal paper in the journal *Nature*, claiming that the mtDNA found in representatives of all the living peoples of the world could be traced to a common ancestor who lived in Africa, approximately 200 000 years ago (Cann et al. 1987). This study also advanced the 'Out of Africa' theory (also referred to as The Recent African Origin or Replacement Model), which claims that there was only one region where there was a complete evolutionary sequence from earlier human ancestors, namely *Homo erectus*, to modern humans, and that region was Africa (Stringer 2002). The 'multiregional' theory, on the other hand, claims that over the last 1–2 million years anatomically modern humans have evolved gradually from their *H. erectus* ancestors in various regions throughout the Old World – Europe, Africa, Asia and Australia – and there the gene flow between archaic and modern humans has ensured that humans have remained one species (Wolpoff et al. 2000). Most evolutionary geneticists and palaeoanthropologists espouse the 'Out of Africa' theory of human evolution.

A very convincing argument in support of the 'Out of Africa' theory was made when it became possible to extract DNA from a Neandertal fossil specimen dated to about 40 000 years ago, and to derive an mtDNA sequence from it. Comparison of the mtDNA sequence of Neandertal man with that from over 3 000 modern humans revealed that there were, on average, twenty-eight differences in the segment of about four hundred base pairs of mtDNA between the Neandertal and humans (Krings et al. 1997, 2000). These findings suggested that mtDNA in modern humans and Neandertals diverged from a common ancestral type over 650 000 years ago. More recently, two additional Neandertal specimens – the Mezmaiskaya specimen from the northern Caucasus, and a specimen from the Vindija Cave in Croatia – were analysed and confirmed the earlier findings (Ovchinnikov et al. 2000). Interbreeding between Neandertals and modern humans cannot be excluded, but the findings show that Neandertals did not contribute mtDNA to the contemporary human gene pool.

One of the most significant findings to emerge from DNA studies is that the gene pool in living peoples outside of Africa is derived from an ancestral population that experienced a severe reduction in population size, a 'bottleneck', at some time in the past, followed by a population expansion (Ingman et al. 2000). This bottleneck and expansion is presumed to have occurred when a branch of the early modern human population in Africa split off to form a small sub-population, which then expanded in size as it spread out to colonise Eurasia. The result of this bottleneck effect is that the gene pool found in African populations is much more diverse, with a larger number of ancestors, when compared with that found in non-African populations (Ingman et al. 2000).

The distribution of mtDNA types among populations from different regions of the world is consistent with the 'Out of Africa' theory. The human mtDNA tree shows that the deepest branches of the tree, that is, branches most closely associated with the common trunk (the equivalent of the most recent common ancestor), are branches L1, L2 and L3, which are found among African populations (Forster 2004; Soodyall, unpublished). In addition, all branches of the tree found outside of Africa can be derived from L3 and therefore must have evolved from one of the more ancient African types (Forster 2004).

Himla Soodyall, who joined the Department of Human Genetics at the SAIMR/University of the Witwatersrand in 1987, used mtDNA to examine the genetic affinities of populations in southern Africa for her doctoral studies under the mentorship of Trefor Jenkins. Many of the genetic relationships observed using the classical

GLOBAL PATTERNS OF Y CHROMOSOME VARIATION

The human sex chromosomes look very different from each other, but far back in the evolutionary process they were once a pair of non-sex chromosomes, i.e. homologous autosomes. The process of divergence was initiated when one of them acquired a male sex-determining function early in mammalian evolution, followed by a process which prevented them from exchanging genetic material (i.e. recombining) during the production of sperm, and this introduced differences between the two chromosomes. The Y chromosome, which is transmitted exclusively from father to sons, is found in the nucleus of all males. In addition to being inherited exclusively from the male parent and comprising about 50 million base pairs in length, it does not recombine. This non-recombining region of the Y chromosome, abbreviated NRY, has become an extremely important tool since the discovery in 1985 of the first genetic polymorphisms on the NRY. It is used in a variety of areas of particular interest: anthropology, forensics, medical genetics, genealogical reconstruction, molecular archaeology, non-human primate genetics, and human evolutionary studies.

The patterns of Y chromosome variation found in representatives of the world's populations are divided into 18 major arrangements, known as clades (with further subdivision in some instances), designated A to R before adding minor patterns of variations. The first two branches (A and B) lead exclusively to African lineages; haplogroups C and D are largely confined to East Asia; E to Africa, West Asia and Europe, whilst the other branches contain both African and non-African branches. The time to the most recent common ancestor for the entire Y chromosome phylogeny is estimated to be about 80 000 years before present.

marker systems were confirmed, but the mtDNA studies permitted in addition an assessment of the role of females in shaping the gene pool of living peoples. The inclusion of populations from other countries in the sub-Saharan region has shown that the mtDNA pool of sub-Saharan African populations is composed of L1, L2 and L3 lineages, albeit at different frequencies (Soodyall 1993; Watson et al. 1996; Chen et al. 2000; Salas et al. 2002). Khoisan populations, for example, have a higher frequency of L1 lineages than other populations. More importantly, some of the oldest mtDNA lineages found in living peoples throughout the world are retained in some Khoisan populations (Soodyall 1993; Watson et al. 1996; Chen et al. 2000; Salas et al. 2002; Soodyall, unpublished). It is possible that other populations have lost these mtDNA lineages purely by

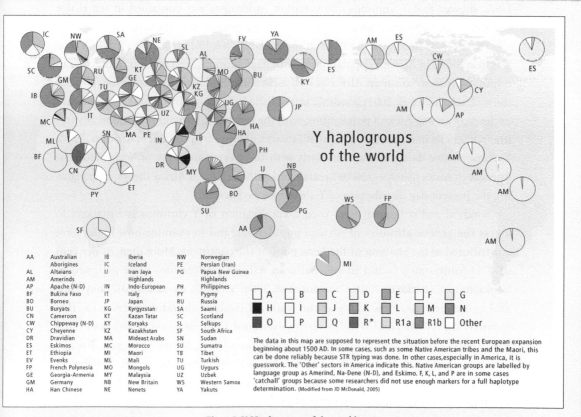

Y haplogroups of the world

AA	Australian	IB	Iberia
	Aborigines	IC	Iceland
AL	Altaians	IJ	Iran Jaya
AM	Amerinds	IN	Indo-European
AP	Apache (N-D)	IT	Italy
BF	Bukina Faso	JP	Japan
BO	Borneo	KT	Kazan Tatar
BU	Buryats	KG	Kyrgyzstan
CN	Cameroon	KY	Koryaks
CW	Chippeway (N-D)	KZ	Kazakhstan
CY	Cheyenne	MA	Mideast Arabs
DR	Dravidian	MC	Morocco
ES	Eskimos	MI	Maori
ET	Ethiopia	ML	Mali
EV	Evenks	MO	Mongols
FP	French Polynesia	MY	Malaysia
GE	Georgia-Armenia	NB	New Britain
GM	Germany	NE	Nenets
HA	Han Chinese		

NW	Norwegian
PE	Persian (Iran)
PG	Papua New Guinea
	Highlands
PH	Philippines
PY	Pygmy
RU	Russia
SA	Saami
SC	Scotland
SL	Selkups
SF	South Africa
SN	Sudan
SU	Sumatra
TB	Tibet
TU	Turkish
UG	Uygurs
UZ	Uzbek
WS	Western Samoa
YA	Yakuts

☐ A ☐ B ◧ C ☐ D ▨ E ☐ F ☐ G
■ H ☐ I ☐ J ▨ K ☐ L ▨ M ▨ N
■ O ☐ P ☐ Q ▨ R* ☐ R1a ▨ R1b ☐ Other

The data in this map are supposed to represent the situation before the recent European expansion beginning about 1500 AD. In some cases, such as some Native American tribes and the Maori, this can be done reliably because STR typing was done. In other cases, especially in America, it is guesswork. The 'Other' sectors in America indicate this. Native American groups are labelled by language group as Amerind, Na-Dene (N-D), and Eskimo. F, K, L, and P are in some cases 'catchall' groups because some researchers did not use enough markers for a full haplotype determination. (Modified from JD McDonald, 2005)

Fig. 4.5 Y Haplogroups of the world

chance or through what are known technically as 'random genetic drift' effects. How-ever, these data strongly argue in favour of the theory that modern humans originated in Africa, and possibly in southern Africa.

The Y chromosome is the paternally inherited equivalent of mtDNA, and is inher-ited from father to son. Most of the Y chromosome is non-recombining – that is, the variation existing in the Y chromosome is brought about by mutation alone. Earlier studies conducted by Mandy Spurdle, a doctoral student contemporary with Himla Soodyall at the SAIMR/University of the Witwatersrand, made use of low-resolution Y chromosome markers to show the genetic relatedness of the various Khoisan and Bantu-speaking groups (Spurdle & Jenkins 1992).

As more useful Y chromosome variation was discovered, Peter Underhill and col-leagues (2000, 2001), working with Luca Cavalli-Sforza at Stanford University in the USA, showed that Y chromosome variation in humans was contained in ten major branches (haplogroups) that had a very specific geographic distribution.

The 'deepest' lineage in the Y chromosome DNA tree, like those in the mtDNA tree, was also found in African populations within haplogroup A, and was found in Khoisan populations from southern Africa as well as in some Ethiopian and Sudanese popula-tions (Underhill et al. 2001; Soodyall, unpublished). Haplogroups B and E were found exclusively among African populations, with the remaining haplogroups found at vary-ing frequencies in non-African populations and, at lower frequencies, in Africans. Thus, Y chromosome data are also consistent with the greater antiquity of Y chromosome lineages in Africa (80 000–150 000 years), and with the 'Out of Africa' theory, in explain-ing the present-day distribution of Y chromosome lineages.

Soodyall and colleagues have used a combination of Y chromosome markers to assess the genetic affinities of African populations, and to examine how males have contributed to the shaping of the gene pool of the continent. More than 70 per cent of Y chromosomes found in sub-Saharan African populations were assigned to haplogroup E (Barkhan 2003; Cruciani et al. 2004; Soodyall, unpublished). Tony Lane and colleagues (2002), working at the National Health Laboratory Service (NHLS, formerly the SAIMR) and Wits University, examined Y chromosome variation in seven South African Bantu-speaking groups or populations (Zulu, Xhosa, Shangaan, Tswana, Northern Sotho, Southern Sotho and Venda), and estimated the genetic differences among these groups to be insignificant (1.4 per cent). Expressing this another way, the seven groups share roughly 98.6 per cent of their Y chromosome variation. These find-ings suggest that the studied groups descended from a common ancestral population and have not been isolated for very long (perhaps less than 2 000 years), even though their languages have diverged within that time period.

In the absence of written records, researchers have relied on oral history; historical information; language; anthropological, archaeological and palaeontological data and genetic studies to reconstruct our past. These studies, taken together, confirm that the group of people often referred to collectively as the Khoisan constitute the aboriginal inhabitants of southern Africa.

Southern Africa has received three major immigrations in the last two millennia: the first from people speaking Bantu languages, perhaps in the last 2 000 years; the second, in the last 450 years, from sea-borne European immigrants, who transported 'slaves' from Indonesia, Madagascar and even some from West Africa; and the third, indentured labourers from India, China and the Malay Archipelago in the past 100–150 years. There have been varying degrees of genetic admixture between different southern African populations, resulting in a complex pattern of diversity among the peoples of the sub-continent.

Humans and chimpanzees: Can our genes tell the story of our divergence?

More recently, genomic research has started to unravel some of the ways in which, and the reasons why, humans differ from their closest living relative, the chimpanzee. Although humans and chimps forged separate evolutionary paths some 5 to 6 million years ago, both species differ in their genetic make-up by only just over 1 per cent at the DNA sequence level. One approach that has been adopted to elucidate the genetic footprints of this divergence is to search for genes that reveal signs of natural selection. The assumption is that genes or genomic elements that confer a selective advantage will show more functionally significant molecular changes than selectively neutral genes or regions. A study by Andrew Clark and colleagues (2003) used this approach to identify human genes affected by positive selection, that is, selection that preserves new genetic variants. By comparing 7 645 genes from humans to their chimp and mouse equivalents, Clark and his colleagues identified genes in several functional categories – including developmental processes, reproduction, neurogenesis (development of the central nervous system, including the brain), smell and hearing – that showed strong evidence for positive selection in the human line.

Speech is considered to be a defining characteristic in humans. The forkhead-box P2 (FOX-P2) transcription factor gene has been implicated in speech acquisition and development, and was found to have undergone an unusually high substitution rate in the evolution of the gene following the divergence of the human from the chimp (Enard et al. 2002). The mutations in question are estimated to have become fixed in humans about 200 000 years before the present. Refined speech is thought to have resulted from one of these mutations.

In addition to functional differences, some genetic variants have been implicated in producing morphological changes. A recent study by Stedman and colleagues (2004) demonstrated that the gene encoding the predominant myosin heavy chain (*MYH16*) expressed in certain muscles was inactivated by a frameshift mutation. Loss of this protein is associated with marked size reduction in individual muscle fibres of the masseter (the major masticatory muscle involved in jaw movements). The authors estimated that the mutation arose in the human line about 2.4 million years ago, that is, after the separation of the human and chimpanzee lines. The evolution of the genus *Homo* is associated with the appearance of several defining features or traits; among these is a reduced reliance on powerful masticatory (jaw) muscles as a means of breaking down food and a dramatic increase in cranial capacity (Currie 2004). The dating of the inactivating mutation in the *MYH16* gene to over 2 million years before the present prompted Stedman and colleagues (2004) to speculate that the reduction in the size of the masseter muscle removed an evolutionary constraint on brain expansion, as suggested by anatomy of the muscle attachment relative to the sutures of the skull.

Conclusion

The history of the peoples of southern Africa can be reconstructed using a variety of methods, each having its own strengths and limitations. We are a long way from understanding why we're so different from our closest cousins, but genomic research – comparative genomics, population genetics, gene-expression analysis and medical genetics – has begun to make significant inroads into the complex genetic architecture of human evolution. The study of genetic variation can be considered as another 'tool' for the study of human evolution and history. The 'Cradle of Humankind' and the surrounding areas of southern Africa were inhabited by hunter-gatherer people for tens of thousands of years before the arrival of the Bantu-speaking agro-pastoralists within the past 1 500–2 000 years, and the more recent immigrants from Europe and Asia. Thus, the gene pool of living peoples from the region carries a record of the past interactions of these immigrants with the indigenous inhabitants. Nevertheless, if we were to represent the different genetic lineages (mtDNA or Y chromosome DNA) as branches on a tree, all branches (representing the various lineages found among all living peoples) would be connected via the trunk, with its roots deeply entrenched in Africa, and very likely in southern Africa, possibly close to the 'Cradle of Humankind'. Genetic approaches to addressing questions of anthropological interest (the range of methods referred to as molecular anthropology) have made, and will continue to make, significant contributions towards the unravelling of the evolutionary history of our species.

FOSSIL PLANTS FROM
THE 'CRADLE OF HUMANKIND'

CHAPTER 5

Marion Bamford

The fossil hominids and fauna from the 'Cradle of Humankind' are well known, but it is also important to know the plant environment in which they lived, died and evolved. We can obtain an indication of the environmental conditions by comparing fossil fauna with closely related modern animals whose climate and vegetation preferences we know. Then we assume that the fossil relative lived under the same conditions. Problems arise when the fossil evidence includes a mixed assemblage of animals from different environments or from different times. In addition, the correct identification of the fossil animal is most important.

On many sites we have to rely only on faunal and sedimentological clues to the past vegetation and climate, because no plants have been preserved. Fortunately some sites contain fossil plant material from some of the time intervals. As research continues on these sites, more plant material will probably be recovered. The sites where plant fossils have been found are described below, together with an assessment of these fossils' reliability for palaeoclimate reconstructions.

The taphonomy of plants

Throughout geological time fossils are abundant in some areas, but for the most part the rocks contain either trace fossils (footprints, trackways, tunnels), invertebrate fossils (shells, for example), vertebrates (bones but no soft tissue) or plants. Seldom do two or more of these groups occur together. Yet in life, they most certainly occurred together. The study of the process of transition from a life assemblage to a death assemblage and what happens thereafter is called taphonomy (Efremov 1940; Martín-Closas & Gomez 2004). Clearly, a study of plant fossils is an important aspect of any efforts to reconstruct these processes.

The first observation to be made is that plants occur virtually everywhere on the surface of the earth and in many parts of the sea. Even when plants are alive, leaves fall off (seasonally or less frequently); flowers, fruits and cones fall off; and pollens or spores are dispersed by wind or insects. When the plant dies it usually breaks up even more, separating into leaves, twigs, seeds, flowers, trunk and roots. The different parts of any one plant vary in weight, size, shape and hardness and so are dispersed farther from or closer to the living plant. For example, the trunk and roots of a tree may not move at all. The seeds can be eaten and, if undigested, can be deposited far from the tree by the travelling animal. Dead leaves may blow away in the wind or be trapped in a pond and buried. The pollen is likely to be blown far away or taken away by insects for pollination or food. The separated dead plant material, wherever it happens to be, is usually decomposed by fungi, bacteria and insects, the nutrients returning to the soil and then being recycled by new plants. If something disrupts this natural cycle then there is a chance that the plant parts can be preserved, but that too needs a particular set of conditions.

Firstly, the decomposition cycle must be stopped, or at least slowed down considerably. Fungi and bacteria are ubiquitous but they require certain conditions to thrive, namely sufficient moisture, warmth, oxygen (except the anaerobic ones) and suitable alkalinity or acidity (pH). Such conditions are found in compost heaps or on forest floors. If one or more of the conditions is unsuitable or limiting, the rate of decomposition can be affected; however, there are so many microbes around that it is likely that some of them would be able to survive the altered conditions. Excluding either moisture or oxygen, or both, is probably the most effective way of preventing decomposition.

Secondly, suitable conditions for preservation are required. There are different ways in which plants can be preserved. The most common way is by compression, where leaves, twigs or seeds are buried in fine sediments under water (limiting oxygen) and then compressed by layers of more sediment on top. Sometimes some carbon material or the waxy cuticle remains, but further diagenesis (the changes that occur as decomposing sediment turns to fossil matter) can remove all organic material and leave only an imprint of the leaf in the rock. Another, similar type of preservation can occur in

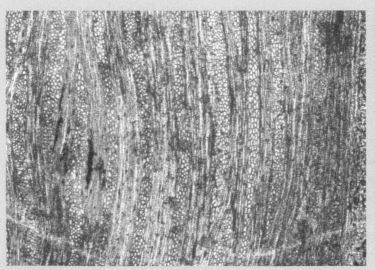

Figs 5.1a & b Thin sections of fossil wood viewed under a microscope

which the outer shape of the plant part is retained, while the inside is replaced by another sediment (mould and cast). Perhaps the most informative type of preservation is petrifaction, where three-dimensional structure is preserved as well as the details of the internal plant cells. Such preservation requires special conditions. Not only must the rate of decomposition be greatly reduced (not stopped completely) but the environment must contain sufficient suitable ions in solution. It is generally thought that the slightly decomposing tissues attract the ions; for example, the silica ions in monosilicic acid, which gradually replace the organic matter with inorganic minerals but retain the same shape as the matter they replace. Calcium carbonate, phosphates and ferric compounds, if present, could replace the organic material; silicification, however, is the most common form taken by this process.

Pollen and spores, because of their small size and extremely tough outer layer (exine), behave somewhat differently and are unaltered after the plant dies. Large numbers of pollen grains or spores are produced by plants. Because of their small size and durability they tend to be transported farther from the parent plant than macroplant remains. These small grains are often incorporated in the fine sediments and preserved, especially if the conditions are dry and slightly acidic. Other plant microfossils are phytoliths, small silica bodies in plant cells, which accumulate in soils or sediments and are characteristic of the parent plant. They are also durable in similar conditions but can be dissolved in acidic or alkaline waters.

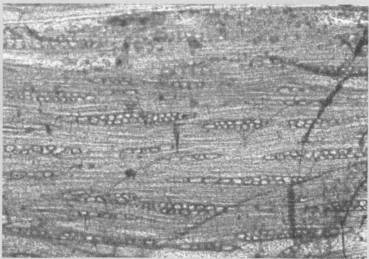

Fig. 5.1c Rays of unidentified wood

Figs 5.2a & b Fossilised plant specimens from older deposits

Fossil plants in the 'Cradle of Humankind'

Bones are often preserved in the solution cavities of the dolomite caves at the 'Cradle of Humankind' because the conditions are suitable and a replacing mineral is available. However, the conditions in the caves are not ideally suited for the preservation of plants. These caves are usually damp, well aerated and very alkaline. The breccias are porous

Figs 5.3a & b A leaf impression and silicified conifer wood

and groundwater percolates through, carrying modern pollen and dust into the caves. Water movement in the vadose zone (the zone above the water table) will destroy pollen grains. Macroplant material is likely to decompose quickly and without a trace.

Yet in Member 4 of the Sterkfontein caves, fossilised wood is abundant. Plants need light for growth, so those that provided the wood did not grow in the caves but must have arrived there by other means. The conditions for accumulation of the wood inside the cave, as well as a decrease in the rate of decomposition and preservation, must have been just right. These conditions were also right for the preservation of the numerous vertebrate and hominid fossils that make this member a most interesting taphonomic setting. We do not yet know all the factors that control preservation, but we can make some interpretations of the environment outside the caves from the wood identifications.

Sterkfontein and Kromdraai

The Sterkfontein fossils are the most substantial evidence of the plant environment currently available. The brief descriptions below indicate the present state of our knowledge of the flora at the different sites at the 'Cradle of Humankind'.

About three-hundred pieces of calcified fossil wood have been recovered from Member 4 at Sterkfontein by Alun Hughes. They are small – less than 5 centimetres long and up to 1–2 centimetres in diameter, but usually smaller – and creamy-coloured inside, with dark staining on the outside. Ground and polished thin sections of the wood, studied under a petrographic microscope, show that there are three types of wood represented (Bamford 1999). The most common one (Fig. 5.1a) has numerous small solitary vessels, well-pitted fibres and tall rays, and has been identified as the liana or woody vine, *Dichapetalum mombuttense* Engl. (Dichapetalaceae). The second wood type (Fig. 5.1b) has small vessels arranged in radial multiples and clusters, and much shorter and narrower rays. It is a shrub, *Anastrabe integerrima* E. Mey. ex Benth. (Scrophulariaceae). The third wood type is very rare and has not been identified (Fig. 5.1c).

The woods were identified by a detailed comparison of their internal anatomy with modern woods, firstly using a computer key and then by checking against the slides of modern material housed in the xylarium in the Musée Royal de l'Afrique Centrale, Tervuren, Belgium. Today the liana *Dichapetalum mombuttense* does not grow in South Africa, only in the Cameroon and the Democratic Republic of the Congo (Prance 1972). It grows over the trees in the gallery forests and, like other lianas, does not have enough strength in its stem to be self-supporting. The shrub *Anastrabe integerrima* grows along forest margins, on dry rocky hillsides, in bushland and along streams from East London to KwaZulu-Natal (Palmer & Pitman 1972).

The occurrence of these two plants in Member 4 times (approximately 2.4 to 2.8 million years ago, although various new dating methods give different ages) indicates that the vegetation was more forested, with at least a subtropical gallery forest along the river and lianas overhanging the trees at the entrance to the cave. The shrub indicates that it was not closed forest but had some open patches. The climate was probably wetter than it is today, as there is no gallery forest along the Blauwkrans River.

The micromammal fauna indicates that there was open grassland at this time (Avery 2001). That may seem contradictory, but one must remember that the wood only represents the vegetation close to the cave entrance. Analyses of other vertebrate fauna indicate a variety of climates (Reed 1997) but are dependent on unquestionable stratigraphic correlations before meaningful comparisons can be made. In East Africa the open-air sites are well dated and can be correlated with other sites because the region was volcanically active. The fine layers of ash were widespread and can be dated radiometrically. In South Africa the sites are mostly in caves. The caves have formed, collapsed and been decalcified or recalcified and talus cones have formed in some of them. As a result the stratigraphy of these caves is very complicated and dating has been based on comparisons of the faunas of South African sites with those of dated East African sites. Here it has been assumed that the faunas are strictly time-controlled but, for various reasons, the South African equivalent faunas could be younger or older than the East African ones. Various new and more precise dating techniques are now being used in the South African caves and this will go a long way in correlating the levels within the caves as well as with the East African sites. (See Chapter 3 for further discussion of these dating techniques.) Research has shown that, prior to the influence of modern humans, the vegetation was much more diverse and complex in a number of areas (Ruddiman 2004; Schwab et al. 2004), so it is difficult for us to imagine what a truly pristine environment looked like. Knowing that human impact on the environment has increased the rates of extinction of plant species, we can assume that it was much more diverse in the past.

The palynology studies (studies of spores and pollen) at Sterkfontein have been less informative. Horowitz (1975) sampled extensively from Member 5 and reported his findings, but a re-analysis by Scott (1982) and Scott and Bonnefille (1986) concluded that the samples were contaminated by the presence of *Pinus,* an exotic imported with the European settlers. The fossil samples from the breccia and travertine of Member 4 produced only rare pollen and unquestionable contaminants (Scott & Bonnefille 1986), so the results cannot be used with any confidence. Carrión and Scott (1999) persisted with the palynology of Sterkfontein and sampled some less porous stalagmite and travertine material from Members 4 and 5. Mostly the numbers of grains were low, but one rich sample had a wide diversity of local extant plant pollen as well as exotics. They concluded that the breccia is very porous and that rainwater, together with modern pollen, percolates through. In response to other workers' claims that contaminants can be recognised and excluded from the interpretation (Cadman & Rayner 1989; Horowitz 1992; Zavada & Cadman 1993), Carrión and Scott (1999) took the much more prudent view that although exotics can be recognised and excluded, all the contaminants of modern, indigenous vegetation cannot.

Recently, phytolith studies have been initiated at Sterkfontein by Louis Scott and Lloyd Rossouw. If the phytoliths have been preserved in sufficiently high numbers and the samples are clearly uncontaminated, then some useful insights into the herbaceous and woody vegetation will be obtained.

No macroplant fossils have been found in the Kromdraai complex of caves, but Scott and Bonnefille (1986) have extracted pollen from Members 2, 3 and 4. Numbers are low, exotic contaminants are common and the species present are very similar to those in the modern soil sample, making it impossible to determine whether or not they are contaminants too. They found what could possibly be fossil *Protea* pollen, because it does not occur in the modern soil sample (Scott & Bonnefille 1986; Carrión & Scott 1999). The frequency is still insufficient for a meaningful interpretation of the past vegetation.

Other cave sites

No macroplants have been recovered at Swartkrans and the pollen record is very scanty (Scott & Bonnefille 1986). At Gladysvale, Berger and Tobias (1994) were the first to mention the fossilised *Phoenix reclinata* seeds discovered in the stony breccia. This common palm tree occurs today in some of the more protected valleys in the area, but as the crown of the tree is destructively harvested by local people who use it for making palm wine, its natural distribution cannot easily be determined. Continued excavations at the site may well produce more macroplants.

At other sites in the Cradle area, and at related sites such as Makapansgat, research has thus far not yielded any further information on the flora that existed during the Plio-Pleistocene time period, that is, 7.00–0.05 million years ago.

What we can conclude about plant fossils at the 'Cradle of Humankind'

Reconstructing the past environment and vegetation is a necessary exercise to help us understand the behaviour and evolution of any animal or plant, and there are two main approaches to achieving this. The first one is based on taxonomy and the assumption that the fossil representative lived in the same manner and environment as its closest modern relative. The accuracy of this method depends on the correct identification of the fossil, and on the assumption that what we see today of the environment has not been disturbed or changed, although it may well have shifted geographically. (Distribution of plants is mostly controlled by the climate – temperature ranges, rainfall – so when a climatic belt shifts northwards, for example, the plants will also slowly shift northwards. Individual plants obviously do not move but the seeds germinating on the north side will be more successful than those germinating to the south.) The latter assumption is, unfortunately, not valid, as humans have had an impact on their environment, mostly a very destructive one that involves removing much of the original vegetation and fauna. At best we can make limited comparisons and reconstructions using the knowledge we have now.

Another approach to reconstructing environments involves making functional analyses of the fauna and flora, and minimising our dependence on taxonomy. For plants, this is based on the assumption that certain physiological adaptations, such as water uptake or reduction of water loss from leaf surfaces, are widespread and reliable indicators of climatic conditions. Some plants may have relict distributions: when a climate belt, and therefore the vegetation belt, has shifted, some plants may be 'left behind' because they have survived the 'new' climate by being protected, for example in a kloof, along a water course or on a termite mound. These plants are not typical of the present

Fig. 5.4 Trees and grasses at Maropeng

vegetation and climate and once fossilised would give a different climate signal from the rest of the vegetation. The whole flora should be considered, thus minimising the effect of the physiologically out-of-place taxa.

Plants and animals are seldom fossilised together, although they coexist. It is therefore difficult to correlate the separated entities because of the taphonomic biases that have selected for various life forms. If researchers are always aware of these factors, they can make what correlations and reconstructions are possible within these acknowledged constraints.

Fig. 5.5 Wild grasses at Maropeng

We know that the flora of the 'Cradle of Humankind' has fluctuated and changed over time. It was most probably much more complex and intricate than the open grassland and woodland that we see today – this, at least, is the conclusion that the faunal and floral indicators as well as taphonomy lead us towards. In the early phases of research in South Africa direct comparisons were made with the East African hominid sites, yet these are open-air sites, not caves. By recognising the taphonomic biases of the different settings, scientists are able to make valid comparisons and so achieve a better understanding of the environment that existed in each place at different times in the past.[1]

Fig. 5.6 Riverine shrubs in the Cradle region

Smuts (far left) and van Riet Lowe (centre)

PART 3

THE EMERGING STONE AGE

Amanda Esterhuysen

As with the emergence of palaeoanthropology, palaeolithic archaeology or the archaeology of the Earlier Stone Age would only truly take hold within the context of the 'evolutionary' or 'prehistoric' fervour of the 1850s. Prior to this, strong creationist opinion generally destroyed, silenced or dismissed studies of stone tools. For example, in the 1600s Isaac de la Peyrère published a study on chipped stones found in the French countryside in which he suggested that primitive humans who lived before Adam had worked on the stones. His book was burnt publicly in 1655 (Howell 1966: 10). During the late 1700s John Frere, an English country squire, discovered a number of stone tools in the same stratified layer of earth as the bones of long-extinct animals. From this he concluded that the tools dated to a remote period, possibly 'beyond that of the present world' (Frere 1800, cited in Proctor 2003: 216). His published findings went unnoticed. As more and more deposits containing tools and extinct animals were exposed and could no longer be ignored, explanations were proffered that were acceptable within a slightly more accommodating biblical framework. Jacques Boucher de Perthes, for instance, collected and organised a number of finds from canal and railway excavations in northern France that provided good stratigraphic information. Convinced that the extinct animals and tools belonged to the same period, he argued that they provided evidence of a human race that had been destroyed by a flood which had occurred prior to the creation of Adam and Eve. Many palaeontologists of the early 1800s were biblical catastrophists of this kind (Trigger 1989: 90–91).

By the 1860s the work of Charles Lyell, Charles Darwin and Thomas Huxley had created a new view of the world and of the antiquity of humans. Although their ideas were still resisted in many quarters, they provided the impetus for the development of Earlier Stone Age studies. John Lubbock, an English banker and naturalist, gave the field of study a name when he divided the Stone Age into earlier and later periods, the Palaeolithic and the Neolithic. However, the initial advances in the Palaeolithic took place in France (Trigger 1989: 94–95). Edouard Lartet recognised a series of phases within the Palaeolithic, which he distinguished according to the animals he found in association with the different tool types. He designated four different periods: Auroch or Bison Age, Reindeer Age, Mammoth Age and Cave Bear Age. This work was taken forward during the 1870s by Gabriel de Mortillet, who argued that the Palaeolithic should be divided according to different tool types that ought to be named after the sites from which they were first described (ibid.). Thus, for example, Cave Bear and Mammoth Age became the Mousterian, because the type-site for these tools was Le Moustier in the Dordogne.

Initially, in the absence of any other evidence, the early tools were seen to be the product of modern humans (*Homo sapiens*) and human development or evolution was seen to be unilinear. It was argued that all humans had progressed through the same developmental stages, characterised by the changes in the lithic sequence. However, this evolutionary way of thinking did not dispel the creationist doctrine of degenerationism, a doctrine that underpinned and continued to influence the 'scientific' or 'biological' theories of racism that emerged in the 1800s and persisted until after the Second World War.

Degenerationism took two forms. Monogenism stated that humans had degenerated to different degrees following the creation of Adam and Eve and in doing so had produced different races (Gould 1992: 39). Climate was seen to play an essential role in determining the extent of this deterioration. Polygenism posited the separate creation of different species of humans (ibid.). The doctrine of degenerationism, paradoxically, offered a means to rationalise why people at different stages of technological development coexisted.

The first published studies of stone tools in South Africa started to appear in 1871. These early descriptions and interpretations, although carried out by amateurs, were clearly informed by the European Palaeolithic studies (Maguire 1997: 44). However, South African prehistory began to define itself and develop independently of Europe between the 1920s and 1950s under the 'powerful patronage' of General Jan Smuts (Shepard 2003: 831), and within a nationalist framework that promoted the development of local science and encouraged scientists to operate independently of Europe. Thus Schlanger (2003: 11) notes that in order to 'capture the distinctiveness and importance of the land's remote past' the emerging archaeologists of the time rejected de Mortillet's classification and began to develop their own South African terminology. By 1925 Smuts, no longer prime minister, had become president of the South African Association for the Advancement of Science. This, however, provided him with the time to indulge and develop his passion for anthropology and prehistory (Shepard 2003: 829) and become an active proponent of the 'South Africanisation of science' (Schlanger 2003: 8).

In 1923 AJH Goodwin, a Cambridge University graduate and the only South African with any formal training in archaeology, took up a post at the University of Cape Town (Deacon 1990). He was not, however, handed full discretionary power to design and shape the prehistory of South Africa at his leisure. Rather, he entered a race against both international and local academics and professionals from a range of different backgrounds to describe and publish typologies and construct the nomenclature of the South African Stone Age. In 1928, for example, Cambridge prehistorian MC Burkitt published *South Africa's Past in Stone and Paint* in which, after only a single visit to South Africa, he was able to 'institute significant comparisons of the South African material with the more systematised products of the European industries' and bring the 'standards of his European experience…to bear on problems of classification and typological relationship' (Forde 1929: 157). On the home front, the battle over the terminological terrain was fought between Goodwin, C van Riet Lowe, a public works engineer, and ECN van Hoepen, a professional palaeontologist and director of the National Museum of Bloemfontein. Goodwin and van Riet Lowe published separate papers on the classification of the Earlier and Later Stone Age in 1925, while van Hoepen presented his classification in a paper read before the South African Association for the Advancement of Science in July 1926 (van Hoepen 1930: 351). In a special meeting held after the July conference the three seemed to come to some agreement:

> It was agreed that the Early Stone Age should be sub-divided into three different type industries [Stellenbosch, Fauresmith, Victoria West], while at the same time it should be clearly understood that, at present, there is no means of dating these three sub-types relative to each other. The Later Stone Age may be divided into three industries [Wilton, Smithfield, Stilbaai]… (van Hoepen 1930: 353, my parentheses).

They seemed to reach consensus over what comprised the Earlier and Later Stone Age (the Middle Stone Age would be described later); however, heated debate ensued with Goodwin and van Riet Lowe taking sides against van Hoepen. Publications were peppered with accusations of unsound methodology, scientific piracy, claims to priority over nomenclature, and reprimands for confusing the amateurs (see for example van Riet Lowe 1929, van Hoepen 1930 and Schlanger 2003).

Schlanger (2003) convincingly argues that the rift between the two parties had more to do with language than with content. He makes the point that by the 1920s Afrikaans was fast becoming the language that defined the Afrikaner nation. It had been introduced into the schooling system and in 1925 was accepted in Parliament at the behest of DF Malan as the second official language alongside English. There was also a movement to make Afrikaans a language of science – what better endeavour than to develop a uniquely South African science in a uniquely South African language? Thus van Hoepen's success in publishing in both English and Afrikaans, his launch of the archaeological journal *Argeologiese Navorsing van die Nasionale Museum* in 1928, and the fact that he began to draw support from the likes of CHTD Heese from Riversdale and other 'professional amateurs' from the University of Stellenbosch, caused some concern amongst English-speaking professionals. Schlanger (2003: 18–21) was able to identify this 'distaste' in a number of letters written from the English quarter:

> Van Hoepen is publishing his stuff in a new series he has invented for the purpose, 'Archeologiese navorsing' written of course in the language of the (theatrical) gods. He is publishing our Smithfield A as his 'Koning' and our Stellenbosch as his Pniel (or Vaal). Lowe is very angry (Goodwin to Burkitt, October 1928, in Schlanger 2003: 18).

> Barnard, Goodwin, van Riet Lowe, Heese, van Hoepen were all at the Kimberley meeting. Heese and van Hoepen I met for the first time; they were doing a little propaganda for afrikaans [*sic*] as a scientific language! We did not try to make martyrs of them (Sir JC Beattie to Burkitt, 22 August 1928, in Schlanger 2003: 18).

> [Heese joined the] ranks of die ware Afrikaans – O ja! O ja! O ja – or het van Hoepen ho[m] geropen, do you dink! (van Riet Lowe to Goodwin, 29 May 1928, in Schlanger 2003: 20).

The language of South African prehistory was also contested in Parliament. Hertzog, a long-time opponent of Smuts, was instrumental in sponsoring the German archaeologist Professor Frobenius instead of a home-grown, English-speaking archaeologist in 1930, while Smuts gave the 'English' the edge by creating a second archaeological post within the Department of the Interior in 1934 for van Riet Lowe, as soon as he had enough power to do so (Schlanger 2003: 21). Smuts went on to support the first meeting of the Pan-African Congress (PAC) in Prehistory in Nairobi 1947, the brainchild of Louis Leakey. He not only flew the South African delegation to the congress in an air-force plane, but also offered to host the second PAC in South Africa (Deacon 1990: 48; Shepard 2003: 832). However, after Smuts died in 1950 the National Party, under none other than Malan, rescinded the invitation to host the second PAC.

While the South African Stone Age developed a distinctive classification of its own, the overall interpretation of the nature and meaning of the technological phases remained largely unchanged until after the Second World War. Researchers were slow to comprehend the real depth of time represented by the African Stone Age, and the implications thereof. Although repeated discoveries of prehumans during the 1930s and 1940s in Africa had planted the idea of an African 'Cradle of Humankind', the technological revolution was still thought to have started outside Africa and to have filtered back down (Deacon 1990: 46). Each technological advancement was considered the work of a new arrival or influence, and the living 'Bushmen' and 'African farmers' were regarded as trapped at different developmental stages of evolution in which (prior to interaction with white colonials) they had begun to deteriorate. This sentiment is clearly expressed in the work of van Riet Lowe:

> We have nothing to show that it (the bored stone) was not introduced into South Africa by the Hottentot, who, incidentally, did not penetrate Rhodesia, and who, here, accelerated the deterioration of the aboriginal Bushman stone culture.
>
> The degeneration of the Bushman is an accepted fact, and his deterioration in the art of working stone is reflected in the coarsening of his implements, in other words, in the deterioration of the true Wilton Industry... (van Riet Lowe 1926 quoted in van Hoepen 1930: 347).

Schlanger (2002: 9) draws attention to Smuts's 1932 publication in the *South African Journal of Science* in which he posits that the European and the Bushman once shared a common ancestor, but that the perceived difference was the result of the one population moving north to a more temperate climate, while the other declined in the harsh African desert. As a result, Smuts says, 'we see in one the leading race of the world, while the other, though still living, has become a mere human fossil, verging to extinction.'

These ideas resonated with the ideals of the Nationalist government. Firstly, they justified colonial expansion, since with shared ancestry Europeans had every right to return to an area they believed they had once occupied. Secondly, the presentation of black people as technologically and ideologically primitive justified separate and unequal education, which had been instituted to ensure that the 'native' did not 'rise above his station in life', as Hendrik Verwoerd stated in his 1954 speech in the South African Senate (Rose & Tunmer 1975: 266). In 1948 the Federasie van Afrikaanse Kultuurvereniginge stated that the native was in a state of 'cultural infancy' and should be guided by the white population, 'most especially those of the Boer nation as the senior trustees of the native', to a Christian way of life (Federasie van Afrikaanse Kultuurverenigings 1948: Article 15).

Ironically, however, while South Africa began to divide the population based on these kinds of ideas the Western world was undergoing a major paradigm shift, which may account for the South African government's growing lack of interest in archaeology and cancellation of the second PAC. The atrocities committed during the Second World War caused many anthropologists to take a firm stand against theories of racism. The newly formed United Nations Educational, Scientific and Cultural Organisation issued a statement in the *New York Times* on 17 July 1950 announcing that a world panel of experts had concluded that race was 'less a biological fact than a social myth and as a myth it has in recent years taken a heavy toll in human lives and suffering'. The war had instigated much debate amongst human scientists about the role of animal instinct in human behaviour. Radiometric dating, another by-product of the war, set the human evolution clock back in time, and during the 1950s stone tools were found for the first time in association with early hominids. By 1957 JT Robinson and CK Brain, working at Sterkfontein, could confidently state that primitive stone tools and australopithecines existed together in the same breccias (Howell 1966: 53), and Mary Leakey discovered a hominid in the oldest layer at Olduvai in association with many primitive tools (Howell 1966: 54). Once these layers were dated it became clear that the Earlier Stone Age pre-dated the European Palaeolithic by over a million years.

For Dart, this provided confirmation for his hypothesis that *Australopithecus* was the missing link between humans and apes and now, in the wake of the war, he could provide a reason for the violent nature of humans. After studying over 7 000 bones from the site of Makapansgat, he concluded that the bones represented the remains of food retrieved from many kills (Howell 1966: 65; Brain 1981: 3). From this he inferred that our ancestor was a 'highly effective hunter, capable of killing the largest and most dangerous animals, and from the high proportion of leg and skull bones in these collections he imagined that *Australopithecus* was a 'headhunter' (Brain 1981: 3). Remains of australopithecines, together with the animal bones, led Dart to hypothesise further that our ancestors displayed cannibalistic tendencies:

> These Makapansgat protomen, like Nimrod long after them, were mighty hunters. They were also callous and brutal. The most shocking specimen was the fractured lower jaw of a 12-year-old son of a manlike ape. The lad had been killed by a violent blow delivered with calculated accuracy on the point of the chin, either by a smashing fist or a club. The bludgeon blow was so vicious that it had shattered the jaw on both sides of the face and knocked out all the front teeth. That dramatic specimen impelled me in 1948 and the seven years following to study further their murderous and cannibalistic way of life (Dart 1956: 325–326).

Dart also posited that australopithecines were skilled tool users who fashioned tools from the bones, teeth and horns left over from their meals (Howell 1966: 59). He called this tool industry the 'osteodontokeratic culture'. His colourful views are expanded on in an interview with CK Brain in Chapter 6. Dart's provocative ideas sparked an interest in the study of bone accumulation agencies amongst other researchers, including Brain. Brain's study of the bone assemblage at Swartkrans and his studies of the habits of living carnivores indicated that carnivores had killed the animals whose bones had subsequently collected in the caves. The presence of australopithecine bones in these assemblages indicated that early australopithecines were less likely to have been hunters and more likely to have been the hunted. Later studies shied away from the violent nature of our ancestors, and bone tools in particular were shown to have been used as digging tools. However, a study of a late australopithecine specimen from Sterkfontein has recently reintroduced the possibility that our early *Homo* ancestor, who we know shared the landscape with a late australopithecine, preyed on this less advanced hominid. Kathy Kuman's views on these findings are also presented in Chapter 6. *Homo* as tool-using regent is also debated back and forth amongst scientists, with animal behaviourists pointing out the tool-using abilities of many animals – especially those of our closest living relative, the chimpanzee.

During the latter part of the 1960s and the 1970s HJ Deacon and J Deacon reassessed the Later Stone Age tool categories of Goodwin and van Riet Lowe, and focused on the climatic and environmental factors that may have influenced changes in the tool industries. In Chapter 7, Lyn Wadley outlines the current debates about early human behaviour based on differing interpretations of Middle Stone Age tools and artefacts. She again highlights the problems inherent in comparing the behaviour of ancient people with that of modern hunter-gatherers. Her discussion of the Later Stone Age captures yet another shift in focus which occurred during the 1990s, in which evidence for symbolism and complex social mechanisms such as gift exchange was sought.

During the 1980s a significant advancement in the understanding of rock art by P Vinnecombe and D Lewis-Williams demonstrated that early Khoisan societies, like all modern human societies, were socially and ideologically complex. In Chapter 8, David Pearce highlights the religious aspect of this art in his discussion of rock engravings found near the 'Cradle of Humankind'.

CHAPTER 6

THE EARLIER STONE AGE

Amanda Esterhuysen

The development of early hominid cultural behaviour is to be found in the first durable traces of material culture revealed by the archaeological record – stone and bone tools. Tools, in general, are objects or devices that we use to do a job more efficiently. We think of them as extensions of our own bodies that allow us to modify or manipulate things in our environment to our benefit. Humans are not the only tool users; animals like chimpanzees, birds and otters have been known to use tools. Humans, however, use tools all the time and depend on them for survival. We also use them in many different ways. The Earlier Stone Age, a period from about 2.5 million years ago to 250 000 years ago, gives us a glimpse into a time when our hominid ancestors began to exhibit limited tool use, not unlike that of the modern chimpanzee, and allows us to trace their progress as in time they became increasingly adaptable, inventive, flexible and creative.

The earliest tools clearly manufactured by hominids date to over 2.5 million years ago, and come from the site of Gona in Ethiopia. These tools show that early hominids (our early ancestors and their relatives) were not only able to select the most suitable raw material for their purposes – fine-grained, homogenous rock – but also knew how to flake it. Many of the bones found with these early tools bear cut marks, which have led scientists to conclude that early hominids were chipping flakes off cobbles to cut meat from animal carcasses. In this way it is thought that these early stone knives helped early hominids to scavenge a high-protein food source in the quantity needed to nourish, and in the long run develop, their brains – the brain being metabolically the most expensive organ in the body. Not all researchers believe that meat was necessarily the only source of high-quality food. They point out that seeds, nuts and various insects like termites also provided a valuable source of protein and fats. Studies of the tools indicate that both sources of food were probably exploited.

The Oldowan Industry

This early stone tools industry has been called the Oldowan Industry, after Olduvai Gorge in Tanzania where these tools were first recognised. The Oldowan technology is fairly consistent across Africa. The tools were mainly simple flakes struck from cobbles (Fig. 6.1). The cores were often only struck a couple of times. It seems the reason for this is that the hominids were manufacturing their tools close to the stone source. With no

Fig. 6.1 Pebble chopper from Olduvai Gorge Bed II

Fig. 6.2 Quartz flakes from Sterkfontein

shortage of raw material, there was no need to overwork individual cores. Occasionally, a core could have been used as a heavy-duty tool.

East Africa has many sites where Oldowan tools are found in primary context – that is, in the same place that they were made or used – and in contexts that can be dated. In South Africa so far there are only two sites with Oldowan artefacts, but these have yet to be securely dated. Kromdraai (B) has an estimated date of 1.9 million years ago, but only has two definite stone tools, while Sterkfontein has a deposit estimated to be about 2.0 to 1.7 million years old, with over 3 000 artefacts. The estimates are based on comparisons with finds from East African sites, which are well dated.

These tools have been washed into underground caves and therefore are no longer in their primary context. The hominids must have repeatedly occupied the shady areas around the cave openings, their discarded tools washing into underground caves during seasonal rains.

Sterkfontein, like Gona, shows that hominids were selecting their rocks carefully from gravels; they were clearly favouring quartz, which was not as common as other rocks but easier to flake. After selecting the raw material they made the tools in one of two ways. The first method, called the bipolar technique, involves taking a rock, placing it between two stones and then hammering down with the top stone. Quartz is brittle; it cracks and pieces flake off. These flakes are very sharp and can be used to butcher and cut meat. The other method is a freehand hard-hammer technique, whereby one rock is hit with another. This once again produces sharp flakes. The main differences we see between regional variants of the Oldowan tools lies in the raw materials used, and how their shapes and sizes influenced the flaking techniques.

The Oldowan Industry is not only consistent across space; it also persists for a long period of time. At sites like Olduvai Gorge in Tanzania and Koobi Fora in Kenya, Oldowan tools remained unchanged until nearly 1.5 million years ago. Oldowan technology thus represents a long period of successful adaptation, which lasted for almost a million years. The earliest tool-making hominids were very successful, and their cultural traditions no doubt were far more elaborate than we will ever understand from their stone tools alone.

There is some debate as to who made these early tools. There are two, or possibly three, hominids on the landscape around Sterkfontein who could have been responsible for the tools. The first is a *Homo* ancestor, the second a late *Australopithecus* and the third *Paranthropus robustus*, a hominid that became extinct. It has been shown that *Paranthropus* would have been capable of making the tools, and *Paranthropus* fossils are indeed found in deposits containing tools. Others counter that because *Paranthropus* was very specialised in its way of life, it would not have needed stone tools to be successful.

Kathy Kuman, one of the archaeologists working at Sterkfontein, believes that *Paranthropus* was too specialised in its diet to have needed stone tools, which are in essence a means of accessing a variety of new food sources. Dr Kuman also argues that it is unlikely that a late australopithecine would have made and used tools because at Sterkfontein an *Australopithecus* cranium, Stw 53 (Fig. 6.3), dating to just over 2 million years ago, bears cut marks. She and Professor Ron Clarke think that it was the victim of the real toolmaker.

This is what Kuman had to say during an interview in 2005 about the Stw 53 specimen that bears cut marks from stone tools:

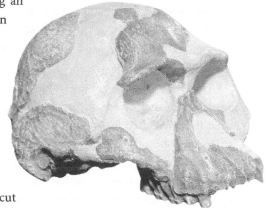

The oldest evidence of tool use in southern Africa comes from Sterkfontein, and it's rather indirect, perhaps even disturbing. In one deposit, which Ron Clarke and I estimate is between 2.0 and 2.5 million years old, we have the cranium of an *Australopithecus*, which bears some cut marks from a stone tool. This is the Stw 53 hominid cranium, which three of our col-

Fig. 6.3 Reconstruction of Stw 53

leagues have now published as bearing cut marks on the cheekbone, which are consistent with its being de-fleshed with a sharp-edged stone tool. Travis Pickering, Tim White and Nick Toth have studied the internal anatomy of these cut marks and their position on the cranium, and they are convinced that the marks are deliberate, and not produced by falling rocks. What convinces me is that we have several hundred other *Australopithecus* fossils from Sterkfontein, which lived between 2.6 and 2.8 million years ago (before we have evidence for artefacts), yet only this one hominid bears cut marks. We also have many tens of thousands of animal bones from those older deposits, as well as bone associated with the StW 53 hominid, and cut marks are absent from them. We have never yet found any damage from falling rocks in the cave deposits which mimics the striations on StW 53, and it seems that Pickering and his colleagues may well be right that this one *Australopithecus* individual surviving into the time of tool-making hominids may have been de-fleshed.

The fact that one hominid may have de-fleshed another may seem disturbing from our modern cultural perspective, but occasional acts of cannibalism have repeatedly been recorded among other primates, like chimpanzees. Our ancestors *were* primates living in the natural world, competing for food sources, and very likely defending their territories against other groups of their own species.

Researchers in the 'Cradle of Humankind' therefore believe that the first tool-making hominids belonged either to an early species of our own genus, *Homo*, or to an immediate ancestor which has yet to be found in South Africa. Early *Homo* was a generalist: from its teeth, which are more the teeth of an omnivore, it is clear that it had a broad-based diet. Early *Homo* was also a survivor – we are testimony to that. It is well known that generalists are survivors in the face of changing environments and competition with other species.

The late Dr Tom Loy of the University of Queensland in Australia analysed the early Sterkfontein artefacts for microscopic traces of organic residues (Fig. 6.4) to determine what they were used for. Interestingly, he found almost equal proportions of plant and animal residues on the Sterkfontein tools. Kathy Kuman suggests that this indicates that early tool use arose not for some singular purpose in the quest for a new food source like meat, but rather as part of a wide-ranging expansion of hominid food-processing activities. She argues further that once tools were invented, they became part of a new hominid lifestyle that brought definite survival advantages to those groups which learnt to depend on them.

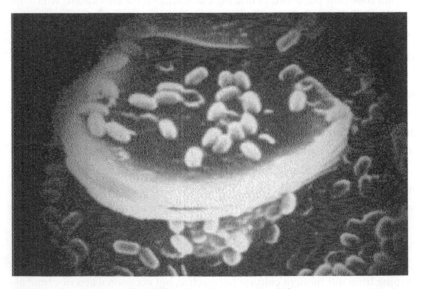

Fig. 6.4 Organic residue on a stone tool

The Acheulean period

The Oldowan continued until about 1.7 million years ago. At about this time a new hominid appeared on the landscape, and with it came more specialised stone tools. These tools were intentionally shaped to carry out specific tasks. For example, these hominids shaped handaxes (Fig. 6.5) – large pointed tools – for hacking and bashing, to remove limbs from animals and to remove marrow from bone. They also made cleavers with sharp, flat cutting edges to carry out more heavy-duty butchering (Fig. 6.6). These specialised tools are called Acheulean tools, named after the French site, St Acheul.

Interestingly, even though the tools are named after a French site, they only appear in Europe after 600 000 years ago. This tells us that the technology originated in Africa and was spread to Europe and Asia through the movement of hominids out of Africa.

Homo ergaster was a much more modern kind of hominid. It had a larger brain size, and a more modern face, body proportions and height. In fact, it had a body very like our own. What is interesting is that *Homo ergaster* appears to have ranged over larger areas of territory. Sites from this time are now found in a greater variety of habitats, including drier, more open settings. In both East and South Africa, there are sites with large numbers of manuports, or rocks that have been transported around the landscape for future use. From this evidence it is argued that over time, *Homo ergaster* became more dependent on tools; it became a habitual tool user.

The middle Acheulean encompasses a period around 1 million years ago when hominids began to show more human traits. The archaeological sites are unfortunately

Fig. 6.5 Typical Acheulean handaxe

Fig. 6.6 Typical Acheulean cleaver

more numerous than the fossils of this period, and the best middle Acheulean collections come from East African sites. Artefacts from Olduvai in Tanzania show that handaxes became more standardised and more symmetrical, and cleavers became more frequent. This reflects the existence of an entrenched cultural tradition characterised by large cutting tools. Like the Oldowan tools, handaxe manufacture does not show regional variation. Differences in biface shape and technique reflect only differences in the type and size of the raw materials from which they are made.

There is some evidence of fire being used during the middle Acheulean in South Africa. This is the earliest evidence for hominids using fire in southern Africa. Dr CK Brain found a consistent build-up of burnt bone in Member 3 at Swartkrans. Every square and level that he excavated in Member 3 had burnt bone; in other words, there was a 6-metre build-up of deposit filled with burnt bone. He posits that the reason for the sudden influx of burnt bone is that hominids sitting near the entrance of the cave were tending fires. The burnt bone washed into the cave later. He feels it is unlikely that they were able to make fire, and that they captured fire from natural grass fires. A study of the bones has shown that they were burnt at a very high temperature for a long time. Bones burnt in an incidental veld fire are not burnt to this degree.

Archaeologists argue that the period beginning 600 000 years ago marks the development of a later Acheulean. From this time until about 250 000 years ago, bifaces show more refinement and are more elegant than those of the middle Acheulean. It is only in the final phase of the Earlier Stone Age that stone tools begin to show some regional specialisation. Initially archaeologists interpreted these specialisations as adaptations to environment. Small, refined handaxes were said to be a grasslands adaptation, while heavy-duty tools were said to reflect the first occupation of more tropical, wooded areas.

Today archaeologists see the issue as being more complex, with a possible social dimension; whatever the reason, it marks yet another advancement in hominid behaviour.

In concluding this short overview of research on the Earlier Stone Age use of tools, we can say that the technological progression from Oldowan flaked cores to the refinement of Acheulean tools provides crucial information about the changes in cognition and behaviour that preceded the emergence of our own species, *Homo sapiens*, about 150 000 years ago. The various studies that have been carried out also plot the progression of ideas about the nature of early hominid behaviour. Past and ongoing research work at the 'Cradle of Humankind' has generated many central hypotheses that have directed and guided archaeological research in the area, and will continue to do so.

Fig. 6.7 Dart at Makapansgat

EVOLVING INTERPRETATIONS OF BONE TOOLS

Bone tools have also been found at sites like Swartkrans, and it has been suggested that they were used variously to dig up tubers and more recently to dig into termite mounds. As discussed in the Introduction to Part 2, Raymond Dart started the debate about bone tools and their possible functions in the 1950s. Mary Leakey notes in her autobiography:

> I was also anxious to see certain other material in which Raymond Dart was an ardent believer, which he had named (with tongue at least partly in cheek) the 'osteodontokeratic culture' of the australopithecines – or in less classical language, fragments of bone, tooth and horn which he was convinced they had used as simple tools. In an impassioned address he had put forward the theory at the Kinshasa Pan-African Congress in 1959, the same one to which Louis and I had taken Zinj. Dart's speech was hugely enjoyed by the audience, but it left them unconvinced. Now I needed to see the objects for myself...I came away unconvinced by Raymond Dart's supposed bone tools, which were not like our few but rather more impressive worked or deliberately broken bones from Olduvai... (Leakey 1984: 136–137).

Dart's ideas were based on his belief that the early hominids were fairly advanced hunters, whose tools were used to carry out fairly bloody business. During an interview CK (Bob) Brain revealed more about the person behind the ideas, and discussed how he (Bob) had disproved some of the early imaginings put forward by Dart.

Interviewer: How did you come to work at Swartkrans?

Bob Brain: ...my original intention of getting a very large sample together from Swartkrans was entirely inspired by Raymond Dart. He actually provoked people like myself into doing further work. You know at the Makapan lime-works cave...there's such an enormous accumulation of fossil bones, hundreds and hundreds of thousands of fossil bones and among these some early hominids. And Raymond Dart really did a very pioneering bit of work in the 50s and the 60s, this was long before there was a discipline for interpreting bone accumulations in African caves. He didn't mind that, he plunged in, he prepared a sample of 7 000 bones from Makapan and he then drew up his own concept of how those bones got there and what their presence meant and he concluded that the bones had been brought to that cave by the ape-man, by the *Australopithecus*, that these people were mighty hunters, that they could kill virtually any animal in that environment and they brought back certain bones to the cave which they would then use as tools and weapons. He developed his idea of the osteodontokeratic culture – bone, horn, tooth culture of *Australopithecus* – and he did this in a most dramatic [way]; his writing was highly dramatic. He would talk about the blood-bespattered archives of humanity for instance, and how when you only found skulls for instance of baboons and of ape-men, this was because those early people were headhunters and they were professional decapitators. And these hunters, Dart maintained, were not only capable of killing whatever animals they chose but, as he put it, they slaked their ravenous thirst on the hot-blooded victims... I used to ask him in those days, in the 60s, why in serious scientific writing he used such dramatic prose and he didn't hesitate for a moment. He said to me, 'That'll get them talking,' and it certainly did. It inspired people like Robert Ardrey, the American dramatist who then wrote four books on this topic starting with *African Genesis*, and then the others followed, and it certainly provoked me into wanting to know if bone accumulations in other caves in South Africa told the same story...

Fig. 6.7 Worked Bone

So I came to Swartkrans and started working here. I didn't know how long it was going to take or what a sweat it was going to be to understand the complexity, that we got this enormous collection together. And it immediately became apparent to me, pretty soon after that, that we were not dealing here with ape-men who were mighty hunters at all. We were dealing with an ape-man who actually was the hunted rather than the hunter. And a book that I put together on this topic was called *The Hunter or the Hunted*, and what this was, really, was an introduction to a new discipline that we call cave taphonomy. Taphonomy meaning, literally, the studying of tombs, what bone accumulation can tell us about the life and death [of] the people and the animals that contributed to it. And it soon became apparent to me that Dart's highly imaginative and highly dramatic explanation at Makapan could not be applied here. But we were dealing here with a situation dominated by the large carnivores; the leopards, the sabre-tooth cats, were the dominant animals, the most powerful animals at the time and...these hominids, the early people and particularly the ape-man, were...the prey of the cats. What we're dealing with here in the fossils was just leftovers after the cats' dinners, and I took all this evidence to Dart as it came out little by little, and I never ever came across anybody with such a generosity of spirit as Raymond Dart had. To begin with he was taken aback, because I was undermining his beloved concept, but within ten minutes he became more and more enthusiastic and he said at last we are starting to get new insight on what was going on, and he was extremely generous to me. He helped me in whatever way he could, the first time I took him evidence that virtually destroyed some of his concepts, he nominated me for an award and he was absolutely delighted...

You know, one of the interesting things was that we found bones that appeared to have been used as tools here. And I was very excited about these, they seemed to have been used for digging, and we asked Raymond Dart and his wife to come over to the museum, he was already over ninety and his eyesight wasn't very good, but he felt each point, each bone point and then he said, 'What do you think these are being used for?' and I said, 'I'm almost certain they were being used for digging in the ground,' and he just sat back in his chair in absolute horror and he said, 'That is the most unromantic explanation I've heard of in my life,' and he took one of the points, one of the long pieces, he stuck it here in my ribs and he said, 'Brain, I could run you through with this.' But he was delighted that somebody had found something that was really close to his heart and had interpreted it in a different way. That didn't matter to him at all, as long as the subject proceeded.

Recently, a systematic study of the bone tools demonstrated that the bone tools retrieved by Brain were used to dig into termite mounds. Lucinda Backwell carried out a comparison of the striation marks on modern tools, which she had employed in a number of different ways, with those detected on the ancient bone tools (D'Errico, Backwell & Berger 2001). The marks on the ancient tools most closely matched those that she had used to dig into termite mounds.

Fig. 6.8 Excavation site at Swartkrans

CHAPTER 7

THE MIDDLE STONE AGE AND LATER STONE AGE

Lyn Wadley

The Middle Stone Age

The African Middle Stone Age (MSA) is broadly comparable to the European Middle Palaeolithic in time and in the types of tools that were produced. However, the MSA differs from the European Middle Palaeolithic in that its tools were made and its subsistence strategies were planned by a different hominid from that living in Europe. The European Middle Palaeolithic was a Neandertal product and no Neandertals ever lived in Africa. Neandertals appear to have been a genetic dead end and they were replaced in Europe by anatomically modern humans (*Homo sapiens sapiens*) about 40 000 years ago. Africa is almost certainly the ancestral home of these modern humans.

There is no consensus definition of the MSA. For one group of archaeologists, the MSA is a time-related sequence of approximately a quarter of a million years, ending about 25 000 years ago. Another group of archaeologists thinks about the MSA as a package of technologies. We will look at the contents of this package later in this chapter. A third group equates the MSA with anatomically modern humans and archaic humans (for example, 'Kabwe Man', *Homo rhodesiensis*). A fourth group of archaeologists believes that the early *H. sapiens* people not only looked similar to us, but that they behaved in a relatively modern (hunter-gatherer) way: they created and used symbols, spoke a language, hunted more often than they scavenged game, controlled fire, and lived in caves as well as open campsites. Other groups of archaeologists accept various combinations of these four hypotheses.

What is Middle Stone Age technology?

MSA stone tool technology lacks the large handaxes and cleavers of the Earlier Stone Age (ESA) and, generally speaking, MSA tools are smaller than those of the ESA. Most MSA flake blanks (thin stone pieces deliberately removed from a chunk of rock) are in the region of 40–100 millimetres, but regional variety of rock types affects the final tool size. The majority of MSA stone flakes are irregular and may have been waste products from

Fig. 7.1 An assortment of MSA stone flakes

knapping (Fig. 7.1). However, some MSA tools are made from prepared cores that are pre-shaped in such a way that they produce flakes of standardised size and shape. Depending on the type of preparation used, different end products are created. Long, parallel-sided flakes called blades comprise one popular MSA tool class. A blade is defined as a flake that has a length more than twice its width. Blades usually have sharp cutting edges and they are sometimes secondarily shaped to form a variety of knives, pointed tools or scrapers. Another common MSA product is the triangular flake, which was probably deliberately shaped to take advantage of its point. When the bases of these triangular points are thinned, they can be hafted (mounted onto shafts or handles) and used as spearheads. Sometimes the points are formed or sharpened by secondary trimming of the flakes' edges. Changing fashions gave rise to a variety of point styles through time.

In recent years, edge-wear and residue analyses have confirmed that many MSA points were once hafted, using resin-based mastic (a type of Stone Age superglue) and bindings of plant fibre or leather. Animal residues, such as blood and collagen, often occur on the tips of the points, which are assumed to have been spearheads, but occasional plant residues suggest that these weapons could have been multi-purpose tools.

A wide range of coarse- and fine-grained rock types was used for making MSA stone tools. Sometimes these rocks were transported considerable distances, presumably in bags or other containers. On such occasions, the stone knappers usually carried out part of the manufacturing process at the rock source. Thus, sites that have tools made from 'exotic' rocks generally lack the full component of the knapping process; they typically contain mostly finished products like flakes and formal tools.

MSA points, blades, triangular flakes and various flaking products are to be found in the Cradle, but little work has been conducted on this relatively recent set of industries, perhaps because of the great wealth of ancient fossils and ESA material culture in the area. MSA tools have been found at Sterkfontein, Swartkrans, Coopers D and Plover's Lake 2. New excavations at Swartkrans by University of the Witwatersrand staff and students, beginning in April 2005, are specifically designed to uncover MSA material; these excavations will probably reveal much local information that was previously unknown.

Between about 70 000 and 60 000 years ago, some groups of people in South Africa began to make a new type of tool that began with long, thin blades. The toolmakers trimmed the edges of the blades with stone or wooden hammers. Segment- and trapeze-shaped tools of about 40 millimetres in length were produced by blunting (backing) one lateral of a blade and leaving the second lateral as a sharp cutting edge. The backed tools may represent a new style of spear armature; certainly residue analysis suggests that these tools were mounted on shafts. No MSA backed tools have yet been found in the Cradle, but there is no reason why they should not be found in the future.

Bone seems to have been used for tools even in the ESA; here bone tools from Swartkrans, in the Cradle, have the type of scratch marks that experimental work suggests could have been used to remove termites from their mounds. However, ESA bone tools were shaped through use, whereas MSA tools were deliberately shaped by people who used them. Outside the Cradle, some South African MSA sites as old as 77 000 years (like Blombos and Klasies River) have preserved bone tools that were ground into a smooth shape. Some bone points may have provided alternatives to stone awls and been used as weapons. In contrast to a stone point, a bone point is incapable of inflicting a deadly blow on a large antelope unless it is merely the vehicle for carrying poison. It is extraordinary to think that people who lived 77 000 years ago had the technical knowledge to make not only strong glues, but also poisons that would fell an antelope without making its flesh toxic.

Middle Stone Age subsistence

From the onset of the MSA, people lived in a wide range of habitats in many different types of campsite, presumably because they had better control over the environment than their antecedents. Sometimes MSA open-air camps were close to pans, lakes or rivers, though their occupants were not as dependent on nearby supplies of water as their ESA counterparts had been. This independence suggests that they had water containers that could have been made of skin or ostrich eggshell. Many rock shelters and caves were also home to people in the MSA, whereas there is only sparse and late evidence for ESA use of such accommodation. Prior to control of fire, rock shelters and caves would have been too dangerous for human habitation; they would have been predator lairs.

Well-preserved MSA sites contain many circular, ashy fireplaces, proving beyond doubt that these people used and reproduced fire. Smashed, burnt bone and charred plant material in and around fireplaces at well-preserved sites show that people cooked some of their food. In cave and shelter sites, which are cultural traps (because they have accumulated tools and rubbish that built up because of repeated occupation by generations of people), it is possible to see changes through time in hunted animal species and gathered plant species. To some extent these reflect adaptations to environmental variation that was stimulated by glacial and interglacial phases.

South Africa was not covered by ice during glacial phases, but glacials inflicted colder than present temperatures and the cold was sometimes accompanied by dry periods. The last glacial occurred between approximately 10 000 and 75 000 years ago and therefore spanned parts of the Later Stone Age (LSA) and MSA. We presently live in an interglacial phase and the last interglacial was between about 75 000 and 130 000 years ago, a period entirely associated with the MSA. Glacial phases are thought to have been associated with lower plant and animal diversity than interglacial phases. Grasslands seem to have expanded at the expense of woodland during glacials, and hunters in the colder phases of the MSA would have had far more large grazing animals to hunt than their interglacial counterparts.

People living in the MSA were true hunters; one of the reasons that we know this is because stone spear tips are sometimes found deeply embedded in bones of animals that have been archaeologically recovered. In the ESA, people were mostly opportunistic scavengers who preyed on the misfortune of drowned or crippled game, or took a share of another predator's kill. Consequently, ESA sites have a high proportion of large, often dangerous game. In the MSA the pattern is different, even though it is highly likely that people still scavenged when the opportunity arose. Hunters in the MSA generally targeted medium-sized animals and, depending on the region being exploited, eland, hartebeest, wildebeest or zebra were typical prey. At MSA living sites, bone is generally smashed to small splinters, presumably because people attempted to extract every scrap of marrow before discarding the bone. At Plover's Lake 2 in the Cradle area, there are bones of many animals. Some of these may have been the prey of humans because they have cut marks on them.

At many South African MSA sites there are charred seeds and underground corms; these suggest that plant food was an important dietary component. In the Cradle area the plant food that is most likely to have been indispensable is the carbohydrate-rich, underground rootstock of *Hypoxis* (African potato), which is a dominant plant in sour grassland. Modern medical science has shown that the African potato boosts the immune system; possibly our African forebears prospered, in part, because of the combined medicinal and carbohydrate-rich properties of this plant.

Middle Stone Age social behaviour
It is unwise to compare the behaviour of ancient people with that of modern hunter-gatherers, not least because such comparisons insultingly imply that modern hunter-gatherers are 'primitive' fossils. In reality, we have no idea how people's behaviour in the MSA compared with that of modern hunter-gatherers. What we can suggest is that these people created and used some form of symbolism as far back as 77 000 years ago. Future research may push this threshold further into the past. At Blombos Cave on the Western Cape coast, people were making shell beads by 77 000 years ago. The Blombos beads have grooves on them; these show that the beads were suspended either on clothing or

on people's bodies. The wearing of ornaments is an indication that people were conveying individual or group identity. This type of behaviour implies symbolic thought. The
use of symbolism is one of the key components of modern behaviour and it suggests
that by at least 77 000 years ago people had the capacity to think in similar ways to us.
At the same time, at Blombos, people were engraving pieces of ochre with geometric
patterns and they were using bone tools in addition to long, symmetrical, beautifully
fashioned stone points. Also in the Western Cape, people at Diepkloof Shelter were dec-
orating ostrich eggshell by about 60 000 years ago. No such finds have yet been made in
the Cradle, but this is not surprising given that MSA research there is still in its infancy.
The earliest representational art in Africa was produced in the MSA at Apollo 11,
Namibia, about 27 000 years ago. It is, however, the only artwork of its kind; other painted slabs occur as rarities from about 6 000 years ago.

At Klasies River in the Eastern Cape, human bone dating to about 90 000 years ago
was cut and scraped in the same way that animal bone was treated. This implies a number of potential behaviours; some archaeologists suggest cannibalism, interpersonal
violence or secondary processing of bodies (scraping flesh from bones for ritual purposes). It is possible that ritual was involved because cannibalism is rarely about a need
to satisfy hunger, but the evidence is equivocal.

The Later Stone Age
Later Stone Age sites in South Africa were almost certainly occupied by the descendants
of the early *Homo sapiens sapiens* groups who practised an MSA technology, and in
many cases the same sites have evidence for both techno-complexes. A division of the
MSA and LSA may be more of an archaeological construct than a real schism, although
LSA tool kits (Fig. 7.2) lack characteristic MSA tools, such as points, and the more recent
hunter-gatherers largely abandoned spear hunting in favour of bow-and-arrow hunting
and snaring with trap-lines. As is the case in MSA research, there is debate about attributes that characterise the LSA. For instance, some archaeologists believe that the LSA
began as early as 40 000 years ago, while others consider that, at least in some regions,
MSA technology persisted until almost 20 000 years ago.

The LSA is marked by a number of technological innovations, and by the regular
occurrence of behaviours that seemed precocious in the MSA, where they were found
only rarely. Thus, LSA attributes include bows and arrows, ostrich-eggshell water bottles
(Fig. 7.3a), bored stones (Fig. 7.3b) used for digging-stick weights, grooved stones
for shaping ostrich-eggshell beads or bone points, small stone tools often less than
25 millimetres in length, polished bone tools such as needles, awls, link-shafts and many
bone points, twine made from plant fibre or leather, tortoiseshell bowls, fishing equipment

Fig. 7.2 LSA tools collected from various sites

Fig. 7.3a An ostrich-eggshell water bottle

Fig. 7.3b A bored stone used as a weight

such as hooks and sinkers, leather clothing, bone tools with decoration, high frequencies of ostrich-eggshell beads and other ornaments (Fig. 7.4), artwork, and burial of the dead, often with grave goods. Not all of these attributes appear simultaneously as a package in the LSA, but most had made their appearance by 10 000 years ago.

The numbers of LSA sites increase dramatically by about 10 000 years ago and this suggests that populations were thriving. The demographic escalation may have been influenced, in part, by people's improved exploitation of the land, but it was probably also due to ameliorating climate and environments at the end of the last glacial at about 10 000 years ago. Warmer, wetter climatic conditions would have promoted greater plant and animal diversity than was possible at the height of the Last Glacial Maximum, about 18 000 years ago.

Pottery is a late inclusion at some LSA sites. Some of the pottery was obtained from pastoralists or Iron Age communities in exchange for hunter-gatherer goods and services, but some pottery may have been made by the hunter-gatherers for their own use. The last 2 000 years were a time of interaction between the indigenous hunter-gatherers and the immigrant pastoralists and farmers and many sites contain mixed signatures in the form of material culture items.

Later Stone Age stone tools

LSA stone technology displays volatile stylistic change compared to the slower pace that is evident in the MSA. Rapid change in the style of LSA tool kits is particularly marked in the past 10 000 years. The LSA sequence includes an informal small blade tradition

from about 22 000 to 12 000 years ago, a scraper- and adze-
rich industry between 12 000 and 8 000 years ago, a
backed tool and small scraper industry between
8 000 and 4 000 years ago and a variable set of
industries thereafter. Between 4 000 and 2 000
years ago there are some sites that have only
irregular flakes and few formal tools, and others
that continue the backed tool and small scraper
tradition. From about 2 000 years ago to just a
100 years ago, tools are often informally made,
though small scrapers and/or adzes sometimes continue. Scrapers are thought to be tools that were used
for preparing hide for the manufacture of clothing and
other leather items. Adzes are thought to have been wood-
working tools and they may have been used to make digging
sticks and handles for tools. Backed tools may have been cutting implements, but they may also have been inserts for stone-tipped arrows.

*Fig. 7.4 A string of ostrich-
eggshell beads*

LSA research, like MSA research, has been neglected within the Cradle, although
some LSA tools have been located at Gladysvale. It can, however, be predicted that, as is
the case elsewhere in the Gauteng region, mid-Holocene occupation (6 000 – 4 000 years
ago) will be poorly represented here. Although there is some evidence for mid-Holocene
occupation in the Magaliesberg, populations here and elsewhere in the sour grassland
areas of the highveld appear to have dwindled during what was probably an arid period
that encouraged people to migrate to better-watered regions. Consequently, the backed
tool and small scraper tradition that is common elsewhere in South Africa is unlikely to
be represented in the Cradle.

Later Stone Age subsistence

Hunting with bow and arrow and trapping with snares imply the use of strategies
different from those needed for spear hunting. Spear hunting requires large-group
cooperation to drive game, corner it and deliver the *coup de grâce*. In contrast, arrow
hunting can be conducted by individuals or small groups. The size of some shelters
occupied during the LSA does imply that, for at least part of the LSA, group size may
have been small, or that it varied seasonally. This issue is discussed again shortly.

The humble snare may, surprisingly enough, have been a far more constructive
invention than the bow and arrow. Snares can be set close to home on the paths of animals with fixed home ranges. Rather than requiring people to go *after* meat, the method

Fig. 7.5 A tortoiseshell decorated with ostrich-eggshell beads

brings meat *to* people with little expenditure of effort. Traps can be set by the young and the aged, meaning that they do not have to be dependants. In many LSA sites there are high frequencies of bones from shy, nocturnal, browsing antelope that are particularly susceptible to being caught in snares. The hunting of small antelope may appear less productive than the hunting of large antelope, yet small game does provide regular, reliable meat packages.

Plant food is well represented in LSA sites that have good organic preservation. This may be because people were more dependent on plants than before, but it could be the result of improved organic preservation in these relatively recent sites.

Many LSA sites contain evidence for fishing, not only at the coast, but also from inland rivers and pans. Freshwater molluscs, snails, frogs and crabs were also extensively collected from about 4 000 years ago. Generally it appears that exploitation of all types of plant food, fish and small creatures intensified from about 4 000 years ago.

Stone Age social behaviour
There is far more evidence for the use of symbolism in the LSA than the MSA, and there is also evidence that symbolic beliefs were expressed in a variety of ways in the

Fig. 7.6 Jubilee Shelter

LSA. The large quantities of beads made from ostrich eggshell, marine shell, land snails and bone imply that group and/or individual identities, and thus symbolism, were frequently expressed through the wearing of ornaments (Fig. 7.5). The presence of marine shell ornaments far from the sea suggests that people in the LSA may have exchanged gifts with family members or other associates across long distances. Another possibility, though, is that people travelled extensively and that band membership was seasonally fluid. At certain times of year, small family groups may have aggregated to socialise, exchange gifts, arrange marriages and perform rituals. Evidence for this practice seems to occur in Gauteng, where sites in the bushveld, such as Jubilee Shelter (Fig. 7.6), contain dense artefact assemblages, rich in formal tools and a variety of ornaments. Ornaments were manufactured, not merely lost, at these sites and faunal remains include a wide variety of game that was hunted and/or trapped. Formal spatial patterning is displayed; for example, specific activities appear to have been carried out repeatedly in particular parts of campsites that may have served as aggregation camps. Grassland sites are different in that they contain sparse evidence for occupation; for example, there are small stone tool assemblages associated with few bones of small, collectable creatures. No ornaments or artwork occur and there is no evidence for formal spatial patterning. These may have been dispersal phase camps because it is a universal hunter-gatherer practice to split (disperse) large groups into small family groups for some part of the annual cycle.

It is not really possible to make gender attributions or to make statements about the division of labour in the LSA. We do know, however, that food sharing between the sexes may not have been as common as hunter-gatherer ethnography would have us believe. This conclusion is the result of studies carried out on human skeletons that were buried in Western Cape coastal sites within the last 3 000 years. The bones of these men and women show significant differences in their carbon isotope values. (There are different photosynthetic pathways for summer-rainfall grasses [C4] and leafy vegetation [C3] and animals that eat these different plant types reflect in their bones the isotope values of the consumed plants. Marine animals have isotopic signatures similar to those of the summer-rainfall grasses.) Since we are (as we have been told since childhood) what we eat, it is possible to conclude that the ancient Western Cape men and women had

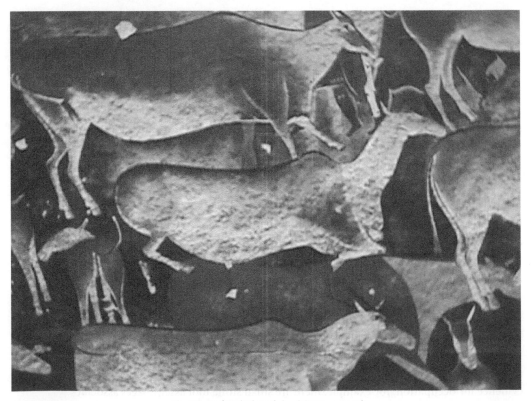

Fig. 7.7 Rock painting of eland (H Pager copy)

different diets. Men's bones were isotopically enriched compared to those of women, suggesting that women were probably eating fruit and vegetables, while the men were consuming larger amounts of protein from marine resources.

It is only during the last 10 000 years of the LSA that there is regular evidence for intentional burial of the dead. People were buried in rock shelters, caves and in the open, and sometimes grave goods were interred with the deceased. Personal ornaments, powdered ochre and headstones were the most frequent grave goods, but painted slabs, ostrich-eggshell water bottles and grindstones were also used. The most elaborate burials seem to have occurred after the period beginning 5 000 years ago. While we shall never know in detail the meaning of the grave goods associated with the deceased, we cannot doubt the symbolic nature of these gifts.

People living in the LSA were accomplished artists and the painted and engraved rock art that they produced throughout southern Africa is a heritage to be treasured (Fig. 7.7). Fine engravings are still to be found in the Magaliesberg, as David Pearce discusses in Chapter 8. The complex friezes in which animals and humans are shown are not merely proof of the artists' skill, but also of their sophisticated religious beliefs. Using Bushman hunter-gatherer ethnography as analogy, it appears that almost all LSA imagery was religious and that the religion was fundamentally shamanistic. In modern hunter-gatherer societies, and probably in LSA societies too, shamans perform an important social role through their curing of social and physical illness. Bushman shamans are reputable rainmakers as well as healers, and so powerful is their control over the elements, that they were called on to make rain for black and white farmers in the nineteenth and even twentieth centuries.

CHAPTER 8

ROCK ENGRAVINGS IN THE
MAGALIESBERG VALLEY

David Pearce

If one wanders across the lower slopes of the Magaliesberg in the late afternoon one may be surprised (and privileged) to see animals peering out from amongst the grass. These animals, though, are rather peculiar. Unlike other game one may see in the area, these are graven in rock. They have been waiting there, largely unchanged, for hundreds, possibly thousands, of years.

Engraved stones such as these are found at dozens of sites scattered across the Magaliesberg valley. Some are cut into the rock using a sharp stone to produce a design in outline. Others are pecked (hammered) into the rock. Still others are made by scraping off the dark patina of the rock to reveal the design in lighter stone beneath. Unlike the better-known rock paintings from other mountainous areas of the country, these engravings depict a large number of animal species. Eland, hartebeest, zebra and rhinoceros are particularly commonly found. Also unlike the rock paintings, the Magaliesberg engravings depict very few human figures.

Fig. 8.1 *An incised engraving of a rhinoceros has many lines cut across its body.*
These were probably produced in the course of rain-control rituals.

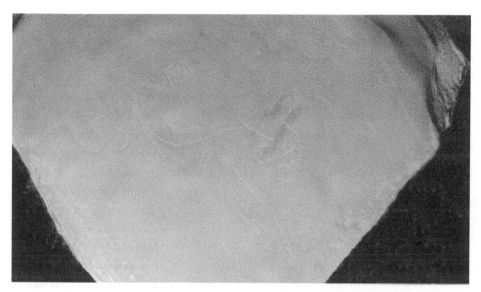

Fig. 8.2 Whilst rhinoceros are commonly depicted in Magaliesberg engravings, human figures are comparatively rare.

Visitors to these sites ask themselves many questions: Who made them? Why are they there? What are they all about? To answer these questions, we need to learn something about the people who made them.

The engravings were made by Later Stone Age people ancestral to the San who lived a hunting and gathering lifestyle. Whilst we have no historical documents relating to these people (there are, though, extensive archaeological data on their economy), we can learn something about why they engraved by examining records of another San group, the /Xam, who lived in the Northern Cape province and who also engraved. Information on the beliefs of these people was collected by Wilhelm Bleek and Lucy Lloyd in the 1870s.

In general terms, the engravings were religious in nature. They related to a complicated belief system in which the cosmos was conceived of as comprising three levels: the everyday human world in which we live, and a spirit world in two parts, the sky above and an underground realm. Ritual specialists (sometimes known as shamans) entered a trance state to travel between levels of the cosmos and to perform several important tasks, such as healing, controlling the movements of game, visiting far-off places and controlling the weather. The engravings played an important part in these religious rites, both as symbols in their own right and as active parts in some rituals.

The most commonly depicted animal in the Magaliesberg is the eland antelope. Throughout southern Africa San people still consider the eland to be the most important animal because it contains the largest quantities of a supernatural potency (*!gi* in the /Xam San language) with which they believe all powerful things to be invested. Shamans use this potency to enter the spirit world and perform their various tasks. One of the reasons the eland is considered to be so special is that, unlike other wild antelope, it contains large quantities of fat. The San say that fat is very potent. Eland are also considered special because males have more fat than females, whereas the opposite is usually the case in other antelope species. Eland are therefore seen as sexually ambivalent.

This idea of sexual ambivalence is depicted in the Magaliesberg in a unique way. An engraving of an eland (Fig. 8.3) has two neck lines, the narrow, straight neck of a female and the large, fat-filled dewlap of a male. It also has two belly lines: the high belly of a male, with a penis sheath, and a lower female belly. It is, therefore, a combination of both male and female characteristics and almost certainly refers to the sexual ambivalence of the animal.

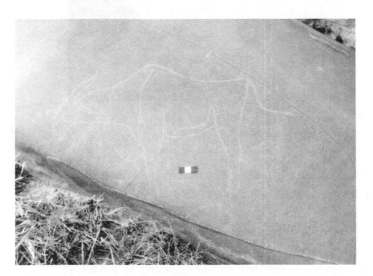

Fig. 8.3 The eland antelope is considered to contain large quantities of supernatural potency. It is also thought of as sexually ambivalent. The double neck and belly lines in this engraving are argued to illustrate this sexual ambivalence.

Another important animal commonly depicted in the Magaliesberg is the rhinoceros. It seems to be particularly related to one aspect of the spirit world: rain. The San conceive of rain as a large quadrupedal animal, often vicious and dangerous. In the Magaliesberg, and some other areas of the country, it seems that the rhinoceros was taken as a model for these rain-animals. To make rain, shamans had to capture a rain-animal from a pool, lead it across the land to where rain was needed, and then kill and cut up the animal. Its blood and, if it was a female, milk would then come down as rain. In general, the engravings of rhinoceros were probably associated with rainmaking. In some special cases, though, they go a step further. Some of the rhinoceros engravings have been cut across the body with a sharp stone (Fig. 8.1). These cuts suggest that the actual engravings were actively used in the ritual of rainmaking: the supernatural cutting of the rain-animal was acted out on the rhinoceros engravings.

Such active interaction between people and engravings and rock is not unique to the rhinoceros depictions. Scattered amongst the representational imagery in the Magaliesberg are many small patches of hammer marks on the rocks. Some of these are focused on other engravings, such as on the neck hump of an eland (Fig. 8.5). This part of the animal is considered to be particularly significant. Hammering this area of the engraving was a significant ritual action; it probably related to the release of supernatural potency.

Isolated patches of hammering away from an engraved image are more enigmatic. Recent research has suggested that the sound produced by hammering rock may have been the reason for the actions that led to these marks. The regular, repetitive percussive sounds made by hammering rock, like the sounds of clapping, chanting and drumming, may well have been an important part of the performance of rituals. The rhythmic sound may have helped shamans to enter a trance, a state that they believed to be the spirit world.

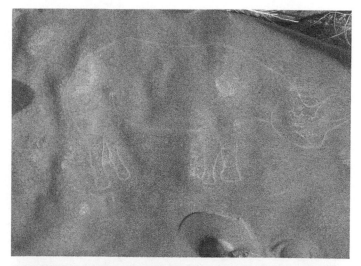

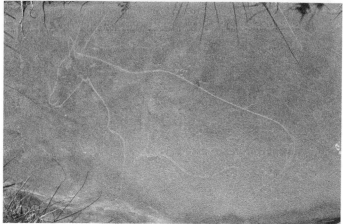

Fig. 8.4 (top) The hammering and cutting of images is an important feature of Magaliesberg engravings.

Fig. 8.5 Hammering on the nuchal hump of an eland engraving probably relates to the release of supernatural potency.

The San believed this supernatural realm to be under or behind the rock face. The process of hammering or cutting through the rock patina when making an engraving was therefore more than a simple, technical process. The animals that we see graven on rock and hidden amongst the grass are not only looking at us from across the centuries, but also from another world.

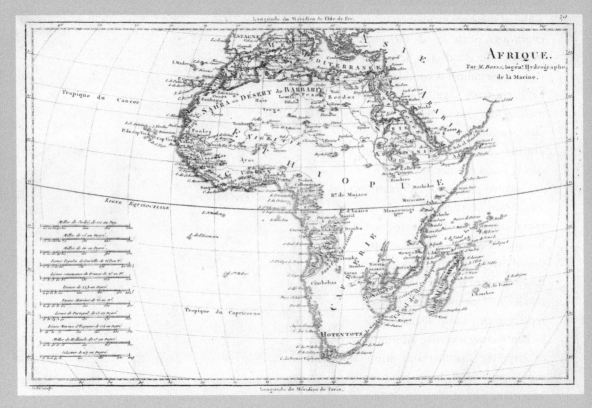

Map of Africa (Ricobert 1750)

PART 4

THE MYTH OF THE VACANT LAND

Philip Bonner

Over the past 1 500 years the Magaliesberg mountains and the Cradle's western border have been host to some of the most densely settled African societies in South Africa. The Cradle fringe boasted both the first Early Iron Age settlement identified in South Africa (dating to around 350 AD) and one of the largest Sotho/Tswana cities which was constructed there in the mid-1700s. Gazing out from the visitors' centre at Maropeng, the eye thus traverses several major sites of early African civilisation and the horizon hides several more. From the early nineteenth century on, African societies and this deep history of African occupation experienced a double obliteration: the first one physical, the second intellectual. The chapters in this section attempt to explain the process by which this occurred.

During the twentieth century most of the interior of South Africa was parcelled out in white farms. White children in white South African schools were taught that this was entirely legitimate, and these claims were justified on the basis of the myth of the empty land. This myth fell into two parts. School texts recorded that the first black South African agricultural communities to immigrate into South Africa crossed the Limpopo River, the subsequent state's northern border, at more or less the same time that van Riebeeck set foot in and colonised the Western Cape in 1652 (Theal cited in Cobbing 1984). Neither Africans nor Europeans, therefore, could assert a prior claim on the land. More sympathetic academic reconstructions based on African oral traditions could only push the date of indigenous colonisation of the interior back two to three hundred years, which scarcely constituted much of a prior right. Most studies, moreover, depicted African societies as being in a state of repeated migration and movement. Like the thinly settled predecessor populations of San and Khoi, therefore, they could not be seen as taming or settling particular tracts of land.

Even these residual assertions were, moreover, entirely extinguished by the accepted depiction of the period of internecine conflict known as the *difaqane* or *mfecane* which immediately preceded the intrusion of the colonial frontier in the form of the Great Trek of 1838. According to this depiction, this period of extreme instability was triggered by the rise of a bloodthirsty Zulu state, which sent victim states in the region now known as KwaZulu-Natal into flight across the Drakensberg into the interior; these groups then, in a famous railway shunting analogy, fell upon other societies deeper in the interior, until the whole landscape was nothing more than a scene of smouldering ruins entirely denuded of population. African peoples now took refuge in a horseshoe-shaped arc of chiefdoms around the edges of the interior – the sites of the subsequent African home-lands. The main exception to this pattern was the Ndebele of Mzilikazi. After taking flight from Zulu attacks in the KwaZulu region, they fled across the Drakensberg into the interior, carrying with them a wave of carnage and destruction. After first centring a conquest state on the Magaliesberg and sweeping the Cradle area clean of human pop-ulation, and then subsequently relocating to the Marico region a little further to the west, they became embroiled with the incoming trekkers in 1838. The trekkers defeated them in a battle as epic and treasured by the Transvaal Afrikaners as the Battle of Blood River (Income), also in 1838, in which the Zulu king Dingane was defeated by Natal Boers. Fleeing north across the Limpopo, Mzilikazi shortly thereafter settled in what is now south-western Zimbabwe, leaving behind a conventionally naked land. Both battles are commemorated in lengthy frieze panels in the Voortrekker Monument outside of Pretoria.

Over the years the *difaqane* was developed as a legitimising myth for the proprieto-rial rights of white settlers to the interior of much of South Africa. GM Theal, one of the founding fathers of colonial history in South Africa, was among the first to elaborate this theme. In 1903 Theal estimated that the loss of life sustained in these convulsions (both in the interior and in KwaZulu-Natal) was 'nearer two million than one' (Theal cited in Cobbing 1984: 4). Since the entire African population of South Africa amounted to only 4 million in 1910, following a lengthy period of sustained population growth, these esti-mates implied total devastation and depopulation of vast swathes of the country – and an open invitation to new settlers. This self-destruction, Theal comfortably contended, rendered insignificant 'the total loss of human life occasioned by all the wars in South Africa in which Europeans have engaged since they first set foot in the country' (Theal cited in Cobbing 1984: 4). Theal's disciple, Eric Walker, shaded in an enormous depop-ulated zone in the early nineteenth-century interior of South Africa in his 1922 *Historical Atlas of South Africa*, variations of which appeared in many subsequent publications (Walker cited in Cobbing 1984: 4–5). These perceptions were parroted in numerous school texts over subsequent generations. Krantz and Trengrove recount how the Voortrekkers crossed the Orange River to enter a land which 'the Zulus had swept clear of human inhabitants' (Krantz & Trengrove cited in Cobbing 1984: 7). Smit and his colleagues explained how refugee African remnants 'spread across South Africa in a horseshoe formation', adding, '[t]his redistribution formed the basis for the present day Bantu homelands, (Smit et al. cited in Cobbing 1984: 8).

The Magaliesberg/Cradle area lay in the centre of this wasted and depopulated zone. The first Boer Republic and the first major town to be established in the Transvaal was at Mooi River (Potchefstroom), within whose boundaries the Magaliesberg/Cradle fell. This is not surprising, given its favoured position close to a juncture of two ecological zones. These environmental factors, however, also ensured that this same favoured ecological niche had been home to innumerable generations of African society stretching back to near the beginning of the last millennium, which gave the lie to the notion of an empty, unsettled, unappropriated land. At least part of this history was wholly unknown until relatively recently. While the few scholars who bothered themselves with African societies recognised that Sotho/Tswana origins in the area could be traced back at least to 1600–1700 (see Simon Hall's citation of Breutz in Chapter 10), no one comprehended the antiquity of African (Bantu-speaking, Iron Age, agricultural) societies in South Africa, and in the Magaliesberg/Cradle area in particular.

Only at the remarkably late date of the early 1970s was an Early Iron Age phase of African agricultural communities clearly identified. Two sites hit the headlines at the same time; the one was exposed at the farm 'Silver Leaves' near Tzaneen, in the northeast of the old Transvaal, the other at Broederstroom 29/72 in the Magaliesberg valley. Both announced the discovery of pottery possessing features similar to the Kwale and Nkope traditions in Kenya and Malawi respectively. The 'Silver Leaves' pottery was recognised as something different by the farm owner, Menno Klapwijk, in mid-1972, when it was exposed during earthwork operations. The Broederstroom pottery was found eroding out of a road on the property of the Leiden Observatory at Broederstroom in the Magaliesberg valley within a remarkably short space of time thereafter, in 1973, and was identified as possessing Early Iron Age features by University of the Witwatersrand archaeologist Revil Mason shortly after that.

Both sites contained charcoal, samples of which were sent for dating at the Council for Scientific and Industrial Research (CSIR) in Pretoria and the University of California in Los Angeles. The dates provided by these two institutions in 1972 or 1973 were the earliest for any Iron Age South African site: AD 230–270 for 'Silver Leaves', AD 460 for Broederstroom. Both sites and dates were first made properly known to the scientific community in the November 1973 issue of the *South African Journal of Science*.

Mason titled his contribution to this journal 'First Iron Age Settlement in South Africa: Broederstroom 24/73' (Mason 1973). Although, strictly speaking, 'Silver Leaves' had been found and identified a few months earlier and also possessed an earlier radiocarbon date, Mason could still claim for Broederstroom the distinction of being the earliest Iron Age settlement in South Africa. Whereas 'Silver Leaves' consisted of two pits, Broederstroom contained thirteen collapsed grain bins within a 15-hectare precinct, a few forges, burnt daga structures, cereal grindstones, thousands of potsherds, teeth of cattle and sheep, and human skeletal remains. In Mason's words, 'Broederstroom represented the earliest intact Iron Age village south of the Sahara known to me and the earliest of cattle farming by negroid people in South Africa.' Mason was also able to recognise that the pottery he had found bore a close resemblance to pottery found by the archaeologist Schofield along the east coast from Natal to East London in the 1930s. In the same issue of the *South African Journal of Science*, archaeologist Tim Maggs confirmed that Schofield's C3 was indeed a hitherto unrecognised Early Iron Age facies, while Mason's colleague at the University of the Witwatersrand, Tim Evers, was now able to announce, in the same issue of the journal, that three sites in the north-eastern Transvaal lowveld (Harmony, Eiland and Langdraai) could now be slotted – slightly later – into an Early Iron Age sequence. Firm radiocarbon dates were, however, still awaited.

A major breakthrough had occurred. Mason, however, insisted that pride of place be given to Broederstroom. 'The search for more data on the Early Iron Age in South Africa,' he declared,

> is now on. There is every prospect that a reasonably complete sequence of Iron Age settlements will be discovered in South Africa perhaps even commencing before the birth of Christ and continuing to the transformations of the nineteenth century. The Magaliesberg valley Iron Age sequence of settlements is now known to extend from Broederstroom 24/77 at ca AD 460, through the Middle Iron Age settlements, such as Olifantspoort 29/72 dated ca AD 1350 on to Late Iron Age settlements such as Olifantspoort 20/71 dated from ca AD 1600 to AD 1820. Environmental conditions in the Magaliesberg valley seem to have favoured a presence of complete settlements in a way more perfect than is common in Africa (Mason 1973).

The Cradle, or at least its doorstep, had been home and witness to yet another epochal phase of human settlement in South Africa. It is difficult to resist posing the question, why did it take so long to establish this? Huffman, who himself conducted a second round of excavations at Broederstroom, and who substantially reinterpreted the site, explores its significance and implications in Chapter 9 of this volume. Here he highlights several key issues raised by Broederstroom and similar settlements. Firstly, he insists that it was the product of physical human migration (rather than the diffusion of ideas), but of immensely earlier antiquity than the atavistic, mindless migrations suggested by previous accounts. Secondly, he argues that it testifies to the arrival of an entirely different kind of human society, which contained the seeds, in the form of the Central Cattle Pattern, of an entirely different order of complexity. Thirdly, he asks, why did a clear identification come so late? The answer he gives is a lack of interest in what can, broadly speaking, be called 'historical' African societies (as opposed to primeval Stone Age hunter-gatherer origins) in colonial Africa; the late development of Iron Age archaeology in East and Central Africa, which largely corresponded to the move towards decolonisation and which lapped into South Africa even later; and the late discovery of carbon-dating techniques and their subsequent application to Iron Age studies (which for the first time offered the possibility of a firm dating for Iron Age settlements). The first carbon-dating laboratory in South Africa was only set up by the CSIR in Pretoria in 1967. Up to this point, without secure dates, the antiquity of the Iron Age could only be guessed at (Inskeep 1970; Huffman [1970] asserted its presence in the northern part of South Africa). Each of these points, of course, suggests a myopia, a lack of desire, which cannot be entirely unrelated to ideological or political reasons, to pay the necessary close and intensive attention to the origins of African societies then settled in South Africa.

The Cradle thus attests to two great African achievements. The first is the Early Iron Age which has just been discussed. The second is the string of massive Sotho/Tswana settlements that stretched from the north-western Cape into today's North West province and Gauteng. This is the subject of Hall's contribution to this section. These settlements, as Hall points out, housed populations of ten to twenty thousand people, were stone-walled and spread over wide stretches of land. To put them into some kind of contemporary perspective, several were larger in number in their heydays than the largest colonial town – Cape Town – at the time. One of the first to be excavated, as Hall observes, lies once again within the perimeter of the Cradle area, at Olifantspoort on the east side of the Magaliesberg; it was excavated by Revil Mason in the early 1970s. The main focus of Hall's study, however, is his current excavation at Marothodi in the vicinity of Sun City.

Whereas Broederstroom over the centuries – or decades – simply sank into the sand and so was lost entirely to view, the Sotho/Tswana cities like Olifantspoort were deliberately obliterated and suppressed. This obliteration occurred in two stages, firstly by violence, secondly by ignoring its historical remains.

The first, violent phase, the *difaqane*, which reduced these cities to smouldering ruins, has been widely misrepresented and misunderstood, notably in white colonial historiography, as a large body of recent historical scholarship has shown. One of the main vectors of violence, as Jane Carruthers notes in her chapter in this section, was provided by outriders of the Cape colonial frontier, especially Khoi and coloureds, who were displaced by trekboer expansion and efforts at their subjugation. Grouped in self-styled Griqua, Kora and other bands – who rode horses and carried rifles, and who sometimes included white brigands like Jan Bloem and Coenraad de Buys in their ranks –

they ranged far into the interior, decades ahead of the Voortrekkers and well in advance of Mzilikazi, spreading violence and dislocation into much of this land. Impacting on densely concentrated Tswana populations, whose trajectory of growth Hall discusses in his chapter, these bands spread first ripples and then waves of conflict into the interior from 1760 on. Many fled and took refuge in defensible spots such as hills or the underground cave complex of Lepalong, or even in trees. They returned to their places of birth as soon as disruption subsided; a few joined Mzilikazi when he arrived in the area. Others, like Mogale's BaPô, who lived on the edge of the Cradle, provided auxiliary forces to the Voortrekkers in their initial battles against Mzilikazi and then returned to settle in their former home. In a number of instances, these groups emigrated once more to escape Boer threats or demands for labour service (see Jane Carruthers's chapter) .

These densely settled communities – the antithesis of the image of an empty land – suffered a second act of obliteration and suppression, this time intellectual, in the twentieth-century era of white supremacy in South Africa. Travellers who passed through the Magaliesberg/Cradle area when Mzilikazi held sway over this territory and shortly thereafter marvelled at the size of the deserted Tswana settlements they encountered, and the scale of the conflict and carnage which had left them empty and in ruins. Generally they attributed these substantial architectural achievements, and the relatively sophisticated political systems that such large settlements implied, to Sotho/Tswana peoples who shortly before had been swept off the land. The chapters by both Hall and Carruthers record some of their reactions. By the early twentieth century, however, any such apprehension or acknowledgement of indigenous accomplishment had been lost. The ruins of the massive Tswana cities which littered the western interior landscape went unremarked or, when unavoidably registered, were attributed to a prior exotic extra-African culture. This, amazingly, was the conclusion of government ethnologist P-L Breutz, who otherwise faithfully collected Sotho/Tswana oral histories from this area documenting a deep history of indigenous occupation that extended back several hundred years (see Hall's chapter), and who marshalled his evidence to support the creation of the apartheid homeland of Bophuthatswana.

Such acts of intellectual marginalisation mirrored policies of territorial exclusion and segregation that were adopted in the early twentieth century (discussed in Part 5). The latter, however, was not easily accomplished and suffered many reverses. When the trekkers entered the Magaliesberg/Cradle area after defeating Mzilikazi, they found wildlife of all sorts in superabundance. This was partly natural, but partly presumably a product of human depopulation. Prior to the expulsion of Mzilikazi hunters were among the first to be drawn to this area, and they continued to comprise a major component of trekker society after they began to settle there (as Jane Carruthers's chapter makes clear). Ivory and skins comprised the major part of Mooi River's economy and exports during the first two or three decades of its existence, and led to a total obliteration of the immense, densely packed herds of game recorded by the likes of Cornwallis Harris (see Carruthers). This constitutes one of the most dismal and formative episodes in the history of South Africa. As herds of elephant and other game were decimated in the Cradle area, hunters moved further afield, especially to the west and the north. Here they entered African-occupied territory and required some measure of African consent and support to carry on the hunt. This enforced a measure of interdependence and self-restraint. As wildlife was obliterated, sections of the Boer Magaliesberg community engaged in agricultural pursuits (see Carruthers). For this they required labour, which was invariably in short supply. Other sections of the Boer community attempted to secure supplies of such labour by taxing local African communities, such as the BaPô, who had returned to the area but who soon resisted and fled such labour demands, or by raiding neighbouring African chiefdoms for children. Some members of the younger generation of the Mooi River Republic made a living out of such raids (see Carruthers) which engendered such instability and violence as to threaten the security and livelihood of more peaceable agricultural Boers. This both divided Boer society and created external African alliances against it. A murky relationship of interdependence and violence, associated with a changing and uneven balance of racial power, characterised the 1840s, 1850s and 1860s. Africans took advantage of this situation by resisting or diluting labour demands, and also by buying back land acquired by right of conquest or by right of 'empty' land from the Boers. This process accelerated from the 1860s through to the latter part of the nineteenth century. The western reaches of the Magaliesberg and areas still further west were one of the principal theatres of such land repurchase in South Africa, much of which was subsequently enclosed within the borders of the bantustan of Bophuthatswana in the 1960s. The BaPô, as Carruthers shows, provided one of the early examples of this kind of repurchase of land adjacent to their ancestral territory of Wolhuterskop in 1862. Only much later – after Union in 1910 – did the South African government interdict this process through the South African Natives Land Act of 1913 (see Part 5). Only then was racial territorial segregation achieved.

CHAPTER 9

THE EARLY IRON AGE AT BROEDERSTROOM AND AROUND THE 'CRADLE OF HUMANKIND'

Thomas N Huffman

About 1 800 years ago Bantu-speaking people brought a new way of life into southern Africa from further north. For the first time, people in this region began to cultivate such crops as sorghum, millets, ground beans and cowpeas, and they herded cattle as well as sheep and goats. Because of the demands of subsistence agriculture, settlements were intended to last a lifetime. As opposed to the temporary camps of pastoralists and hunter-gatherers, farmers lived in permanent settlements consisting of such features as houses, raised grain bins, underground storage pits and animal kraals. Because these early farming people also made their own iron tools, archaeologists call this block of time the Iron Age. For convenience and to mark widespread events, we divide it into three periods: the Early Iron Age (AD 200–900), the Middle Iron Age (AD 900–1300) and the Late Iron Age (AD 1300–1820).

The first two phases of the Iron Age remained totally unknown to South Africans, both black and white, until the early 1970s – an astonishingly late date in the country's history. The first recognition of their existence and the first identified Early Iron Age sites were at the farm 'Silver Leaves' near Tzaneen in the north-eastern Transvaal (now Limpopo province) and at Broederstroom, which lies in the Magaliesberg valley north of the 'Cradle of Humankind'.

As mixed farmers, Early Iron Age people chose to live in broken country where there was sufficient water for domestic use and arable soil that could be cultivated with an iron hoe. With few exceptions, this concern for water and arable land continued until recent times. Generally, the dolomites in the Cradle are unsuitable for subsistence agriculture, and the Cradle is therefore devoid of Early Iron Age sites. The neighbouring Magaliesberg valley, on the other hand, was suitable.

Broederstroom (located at 24°45'E, 27°50'S) is an extensive Early Iron Age settlement in the Magaliesberg valley just north of the Cradle (Fig. 9.1). It was first excavated in the 1970s (Mason 1981, 1986) and then again a decade or so later (Huffman 1990, 1993). These excavations have yielded the remains of settled village life (storage pits, burnt daga houses/grain bins and enormous ceramic vessels), domestic animals (dung, bones and teeth), domestic crops (grindstones, phytoliths and storage facilities) and metallurgy (slag, blowpipes and forge bases), as well as a characteristic ceramic style. Radiocarbon-dated to between AD 350 and 600, it represents the first phase of occupation in the region by Bantu-speaking farmers.

To most South African historians, anthropologists and archaeologists, the finds came as a surprise. A few Iron Age archaeologists working in southern Africa, mostly north of the Limpopo, had hypothesised the presence of an Early Iron Age phase in South Africa (mostly based on similarities between certain ceramic traditions north and south of the Limpopo) but no date of this antiquity had emerged. Now Early Iron Age agricultural communities had been identified 700–800 years earlier than had been commonly thought. The initial political and educational impact of these excavations was thus huge. Subsequently, the Broederstroom site fed into major debates about the character of the Early Iron Age in South Africa. Two debates in particular are worth noting: (1) migration versus diffusion of people through the region; and (2) bridewealth in cattle as a social practice. Both debates involve an understanding of the Bantu language.

The Early Iron Age and the spread of Bantu languages

Bantu languages today are spoken throughout the southern half of Africa. Depending on how one counts dialects and dialect clusters, there are 300 to 800 Bantu languages. All are related by common vocabularies and a distinctive system of noun classes marked by prefixes and what is called a concordial agreement pattern; that is, the noun class determines the form of the verb and so

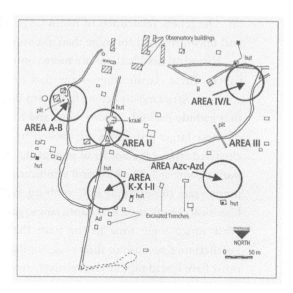

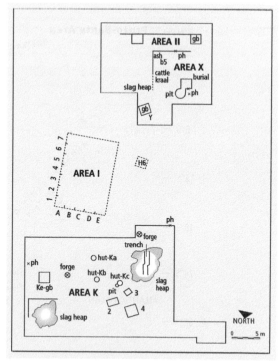

Figs. 9.1a & b Plan of the Early Iron Age settlement at Broederstroom

on. The genetic relatedness of Bantu languages through these features has been known and uncontroversial for more than a century.

Historical classifications are more controversial, but the broad outline is well established. These classifications show that the Bantu languages traditionally spoken in southern Africa mostly belong to Eastern Bantu, which in turn belongs to larger clusters that include Bantoid, Benue-Congo and Niger Congo. Significantly, all other languages of these larger groupings are spoken in West Africa. The most closely related are clustered around the border of present-day Nigeria and Cameroon, the so-called proto-Bantu homeland. The historical significance of this homeland has been recognised for many years (for example in Greenburg 1955). It follows from their distribution and historical classification that Bantu languages were introduced into southern Africa from West Africa some time in the past; they did not evolve within southern Africa. Furthermore, because of their close similarity and wide distribution, Bantu languages must have spread rapidly and relatively recently.

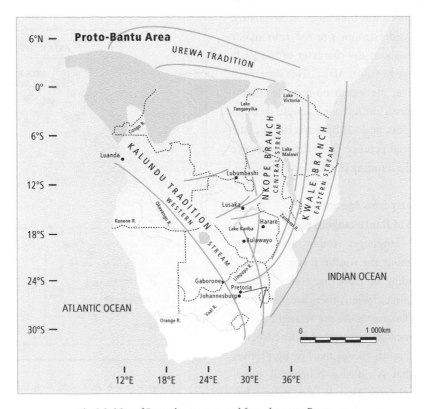

Fig. 9.2 Map of Bantu language spread from the proto-Bantu area

With this last point in mind, Africanists have long been aware of a general correlation between Iron Age mixed farmers and the Bantu language family. For instance, at the time of first European contact only Bantu-speaking people practised mixed farming in southern Africa; and the characteristic Early Iron Age ceramic style spread rapidly and relatively recently. Finally, some stylistic sequences connect historical Bantu-speaking groups with Early Iron Age archaeological cultures.

It is possible to use ceramic style to make these connections for three main reasons. Firstly, because language is the principal vehicle for thinking about the world and transmitting those thoughts to others, there is a vital relationship between world view, language and material culture, such as ceramic style. Secondly, ceramic style is created and learnt by groups of people, and so the transmission of the style must be at least partially accomplished through verbal communication. Thirdly, provided that the makers and users belong to the same stylistic group, it then follows that the distribution of the style must also represent the distribution of a group of people who speak the same language. Note, however, that because languages evolve and diversify, the linguistic scale of the group (that is, whether it is a dialect or a dialect cluster) is an empirical question that must be addressed case by case.

Empirically, Early Iron Age communities throughout eastern and southern Africa share a common ceramic style. The divisions within this common style are sometimes disputed, but everyone who knows the data accepts the broad similarity and general historical relationship that this similarity implies. Indeed, the basic similarity between eastern and southern Africa has been recognised since the 1960s (for example in Posnansky 1961). At the broadest scale, this common style is referred to as the Chifumbaze Complex. The Complex incorporates at least two traditions, Kalundu (also called the Western Stream) and Urewe (the Eastern Stream). Within southern Africa the Urewe Tradition has two branches, Nkope and Kwale (Fig. 9.2). Facies (units of analysis with specific space and time boundaries) within these two branches all share the same range of stylistic types (combinations of profiles, decoration layout and motifs), and their differences are largely limited to the percentages of individual motifs in individual decoration positions. At a broader scale the two traditions share the same basic stylistic structure, but differ in some profiles and motif categories (Huffman 1989).

By tracing backward, phase by phase, the various ceramic styles, it is possible to establish the antiquity of a language group in any one area. In southern Africa, ceramic styles associated with Shona speakers can be traced back from *Great Zimbabwe* to *Happy Rest*, the earliest Kalundu facies in South Africa. *Happy Rest* is the ancestral facies of all Kalundu sequences south of the Limpopo River, and Shona is an Eastern Bantu language. (Shona is the only known surviving language to have evolved directly out of

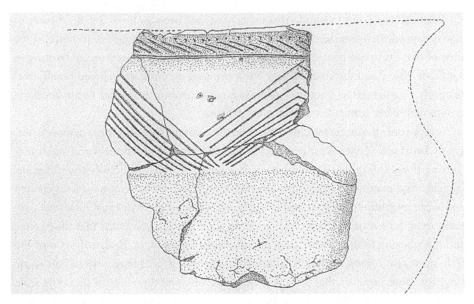

Fig. 9.3a An Mzonjani-style pot from Broederstroom

Fig. 9.3b The Broederstroom pottery sequence

AD 750

Garonga

AD 550

Mzonjani
(Broederstroom)

AD 350

Silver Leaves

the Early Iron Age in southern Africa.) Because of the vital relationship between material culture and language, it follows that the makers of all Kalundu Tradition facies most likely spoke early forms of Eastern Bantu.

A similar link can be made between Eastern Bantu and the Kwale Branch of the Urewe Tradition in East Africa. The Tana facies of Kwale, derived from Kwale itself, dates to between the eighth and tenth centuries AD, and it underlies the walls of coastal Swahili towns. Since Swahili is another Eastern Bantu language, and since no other Early Iron Age facies existed in the area contemporaneously, Tana must represent the indigenous component of proto-Swahili. Because of the close similarities, all Kwale and Nkope facies were most likely produced by early Eastern Bantu speakers. The entire Chifumbaze Complex can therefore be associated with Eastern Bantu.

This conclusion is strengthened by linguistic and ceramic data from outside the Eastern Bantu area. For

Fig. 9.4 Typical grain bin remains at Broederstroom

example, a different ceramic complex characterises early farming people along the fringes of the equatorial forest in Congo and the Democratic Republic of the Congo. This other complex contrasts markedly with Chifumbaze in structure and most stylistic types. This separate association between a separate ceramic complex and separate division of the Bantu-language family strengthens the exclusive association of Chifumbaze with Eastern Bantu.

Originally, I classified Broederstroom pottery (Fig. 9.3a & b) as a facies of Kalundu, but more recent analyses show that it belongs in the Kwale Branch of Urewe. Either way, the Broederstroom people would have spoken some form of Eastern Bantu.

The next line of evidence in support of a migration involves other archaeological remains. Typically, mixed farming settlements contain the burnt remains of daga (a mixture of dung and mud) structures such as houses and grain bins. The remains at Broederstroom clarify the archaeological difference between the two types of structure. Most true huts are built on the ground, and once burnt down, yield flat daga floors exactly like those found in area K (Ka, Kb and Kc) shown in Figure 9.1. Grain bins, in contrast, are usually built on thick daga floors raised on stone supports, and they produce stone-and-floor features such as at Ke (Fig. 9.1) and elsewhere (Fig. 9.4). The thick floors and stone supports are often all that remain of many Broederstroom grain bins, but they are sufficient evidence for cultivation. We believe the crops cultivated included sorghum and millets. Certainly, seed impressions of domestic millet (*Pennisetum* sp.) were found on potsherds at 'Silver Leaves', the name site for the earliest facies of Kwale pottery in South Africa. Carbonised seeds are more plentiful by the seventh century (for example Maggs & Ward 1984), and they include domestic finger millet (*Eleusine corocana*), pearl millet (*Pennisetum typhoides*), sorghum (*Sorghum bicolor*), cowpeas (*Vigna unguiculata*) and ground beans (*Voandzeia subterranea*). Further, phytoliths (microscopic silica formations) from domestic bottle gourds have been found at Broederstroom.

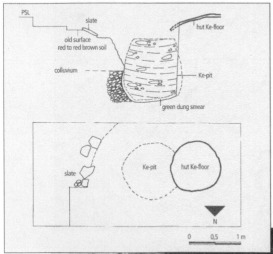

Figs. 9.5a & b Plan and section drawings of typical storage pits at Broederstroom, and photograph of a storage pit at the site

Broederstroom, like many Early Iron Age sites, also incorporated other storage facilities. In addition to raised grain bins, grain may also be stored in underground pits which are smeared with dung and then sealed with a stone. Methane gas from the dung lining kills insects and helps to preserve the produce. If kept dry, the produce is edible for several years and serves as an insurance policy against bad times. After their initial use, these storage pits become rubbish dumps and are of tremendous value to Iron Age research.

The excavations at Broederstroom yielded several such storage pits (Mason 1981; Huffman 1993). A metre-deep pit with a dung lining lay underneath hut Kc, and more were found in Area Azc-Azd (Fig. 9.1). Pit Azzr, for example, was some 2.5 metres below the present surface level, and it contained at least three dung smears (Figs. 9.5a and 9.5b), indicating that it had been used to store grain, then emptied, re-smeared and used again at least twice. As with other pits at Broederstroom and elsewhere, it became a rubbish dump.

These storage pits are important to the migration-versus-diffusion debate. Significantly, the pit fills often contain characteristic pottery together with pole-and-daga fragments, broken grindstones, metal slag and the broken bones of domestic cattle, sheep and goats. Because these items consistently occur together, they show that the Iron Age way of life came to southern Africa as a material-culture package.

Broederstroom and the antiquity of lobola

Another debate involving Broederstroom concerns the nature of early mixed-farming society and the antiquity of lobola – the preference for bridewealth in cattle. In recent times this preference has been a defining characteristic of Eastern Bantu speakers in southern Africa. Among other things, the exchange of cattle for wives underpinned kinship relations and political power. Because of this central role, the antiquity of lobola is an important topic.

Broadly speaking, there have been two schools of thought. In the first, lobola was thought to have evolved in southern Africa a few centuries after mixed farmers entered the region. To some Africanists, the mode of production of Early Iron Age societies shared more similarities than differences with Later Stone Age hunter-gatherers, until a natural increase in cattle herds in the eastern lowveld led to the development of new social relations and lobola (Hall 1986). Before the ninth and tenth centuries, according to this view, cattle were probably not important because the lowveld environment was initially unsuitable. Cattle herds could only increase once coastal forests that sheltered tsetse fly had been cleared by slash-and-burn cultivation. As cattle herds increased over time, cultural attitudes towards them, it was thought, shifted from communal to private ownership.

This first school of thought was based on locational data and faunal remains. For example, early settlement locations throughout East and southern Africa show a preference for broken country with access to water and cultivatable soils: pasturage was not the primary factor. Further, small stock remains regularly outnumber cattle in Early Iron Age faunal samples (as in Plug & Voigt 1985). At Broederstroom, for example, there were remains of only one cow compared to forty-two sheep and goats (Brown in Mason 1981).

The second school of thought, in contrast, held that cattle were already important to Early Iron Age societies and that lobola was practised from the start (Huffman 1990). As part of this argument, the second school questioned the premise that social importance could be inferred from faunal remains alone. It emphasised instead the identification and location of cattle kraals in relation to other features in a settlement.

It is possible to identify cattle kraals, and the dung lining of storage pits, through the analyses of phytoliths. Phytoliths, or plant opal, are microscopic silica formations inside plants such as grasses, sedges and herbs which become incorporated into sediments when the plants decay. The dung of domestic animals, particularly cattle, contains large amounts of phytoliths and other plant matter, and when enclosed, the kraal deposit is richer in this material than elsewhere. With time, these dung deposits

Fig. 9.6 Microphotograph of grass phytoliths typical of Early Iron Age cattle dung

often become grey or khaki-coloured 'ash'. Furthermore, because sheep and goats nibble their food, they break a large proportion of the phytoliths, whereas cattle regularly pass undigested plant matter, leaving clusters of whole phytoliths behind (Fig. 9.6). Thus it is possible to identify early dung deposits and even to distinguish between large and small stock kraals.

Using this procedure on samples from Broederstroom, we could identify at least three kraals in the central area of three different residential units (see Fig. 9.1). The ashy deposit in Area U is a good example: it was an 18 metre oval some 30 centimetres thick. Furthermore, cattle dung filled a storage pit in Area Azzx and this shows that a fourth kraal was present. Yet, despite these data, there was only one cow in the faunal sample. The four kraals so far recognised, and the dung-lined pits, show that the faunal remains seriously under-represent the number of cattle in the settlement. From a biological viewpoint, furthermore, if there was one cow, there had to have been many more in the neighbourhood, perhaps at least one hundred, in order for herds to reproduce. Clearly, faunal samples alone do not accurately reflect cattle numbers.

Once identified, the physical location of these kraals in relation to other features and activity areas becomes important. This is because the spatial organisation of a settlement has social origins, and there are social consequences to spatial organisations. Indeed, the ordering of space is one way of ordering people; so spatial and social organisations are different expressions of the same thing. And both derive from the same world view.

We know from the ethnographic record that one main type of spatial organisation, the Central Cattle Pattern, was associated with Eastern Bantu speakers and lobola (Kuper 1982). Briefly, the Central Cattle

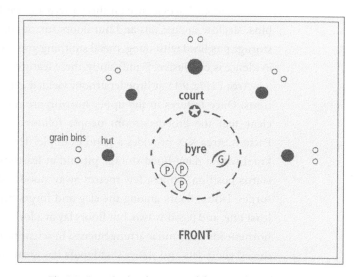

Fig. 9.7 Organisational structure of the Central Cattle Pattern

Pattern is characterised by a male domain in the centre, encompassing cattle kraals where men related by blood and other high-status people were buried, sunken grain pits or raised grain bins for long-term storage, a public smithy, and an assembly area where men resolved disputes and made political decisions. The outer residential zone, the domain of married women, incorporated the households of individual wives with their private sleeping houses, kitchen, grain bins, temporary storage pits, small stock kraals and graves (Fig. 9.7). These outer households were arranged according to a system of seniority expressed through left and right, starting with a 'great hut' built upslope of the court and kraal. For example, in some places today, the first wife lives to the right of the great hut and the second to the left, while in other places the first two wives live on the right and the next two on the left. These small differences are merely variations on the same theme. At a lower scale the same spatial dimension applies to the great hut itself: the central fireplace divides the hut into right–male/left–female space. At right angles is another distinction between front–secular and back–sacred activities that informs behaviour not only in the great hut, but also in the household and the whole settlement.

The ethnographic record indicates that this pattern originated among, and was largely restricted to, Eastern Bantu-speaking people who shared a patrilineal ideology about procreation (one's blood comes from the father), a preference for bridewealth in cattle (lobola), male hereditary leadership and positive beliefs about the role of ancestors in daily life. As a rule, men were associated with cattle and women with agriculture.

Archaeological evidence for this pattern and world view includes the remains of grain bins, shallow storage pits and hut floors surrounding cattle kraals associated with deep storage pits lined with dung, metal smithing and prestige burials. As far as we know, this evidence is conclusive. Significantly, these features occur at Broederstroom.

Area I (Fig. 9.1) at Broederstroom yielded evidence for two Early Iron Age occupations. Once features in the upper horizon are removed from the settlement plan, it is clear that the Broederstroom people followed the principles of the Central Cattle Pattern. Area K-X provides a good example. There the central zone contained a cattle kraal with a dung-lined storage pit and at least three burials, one in an upright high-status position, while a few metres away stood a slag heap and the bases of two iron forges. House floors among the slag and forges belonged to the upper horizon, but at least one, and possibly two, hut floors lay at a lower level outside this central zone. Other homesteads had similar arrangements. In scientific terms, then, Broederstroom has produced good data and shows that lobola has great antiquity in southern Africa.

Broederstroom and the social context of archaeology

Although famous sites such as Great Zimbabwe had generated interest for some decades, Iron Age archaeology only became widespread in the 1960s as the countries of East and southern Africa attained independence. As a result, most Iron Age archaeologists worked in Botswana, Malawi, Zambia and Rhodesia (now Zimbabwe). Even so, there were few Iron Age specialists in a vast area. In South Africa, most archaeologists concentrated on the rich Stone Age deposits, and virtually everyone had been trained to work on the Stone Age.

Relative versus absolute dating is a further factor to consider. Relative dates, such as early, middle and late, are based on stratigraphic relationships, while absolute dates are derived from such techniques as radiocarbon dating. Using relative dating, archaeologists were aware of early African pottery in Limpopo province and KwaZulu-Natal since the 1930s. Accurate dating, however, had to wait thirty years for the development of the radiocarbon method. The establishment of a dating laboratory in Pretoria in the late 1960s greatly facilitated Iron Age research throughout southern Africa. Until then, the antiquity of the Early Iron Age was at best an estimation and at worst of little public interest.

It was as a result of these factors that Broederstroom was only discovered in the early 1970s, and then by accident. According to the archaeologist Francis Thackeray, as a child he found bits of slag while playing outside the astronomical observatory where his father was working. Whatever the exact circumstances, Wits University archaeologist Revil Mason became curious because the pottery was similar to early Nkope in Malawi, and he began an extensive excavation programme. Once the site was radiocarbon dated and its significance realised, Mason had the site proclaimed a national monument. After Mason retired, I returned to search for cattle kraals and to establish the settlement pattern. As discussed earlier, the results of this research were significant to the debates about the origins of the Early Iron Age and the antiquity of lobola.

Now we know that some of the first Bantu-speaking people to enter the sub-continent brought with them a complete 'Iron Age package'. The association between this package and Early Iron Age ceramics is no longer a major issue for debate.

Several years ago, however, some archaeologists rejected this conclusion and proposed instead that farming, herding, metallurgy and the diagnostic ceramic style were introduced separately to an indigenous population (see for example Gramley 1978). Anti-migration hypotheses such as this one have gained popularity since the 1960s for a number of reasons. Indeed, many Africanists still have a deep-seated prejudice against all migration hypotheses because in earlier periods archaeologists used to attribute changes in culture–history sequences everywhere to migration or diffusion. In Africa, this form of explanation was coupled with the belief that 'Bantu hordes' moved across the continent in successive waves, some of them very recently. So in reaction to this, some Africanists now reject all migration hypotheses.

Migration hypotheses, however, cannot be simply dismissed for social reasons; each must be examined case by case. In the case of the Early Iron Age, the physical evidence for settled village life, agriculture, herding and metalworking found together at sites such as Broederstroom completely supports the migration hypothesis, and overwhelmingly disproves the alternative possibility of separate and independent introductions of these practices.

The lobola debate had a somewhat different context. Lobola and the importance of cattle were part of a wider debate about modes of production. In the 1970s and 1980s, Marxism greatly influenced the social sciences in southern Africa, for both social and theoretical reasons. Theoretically, Marxism offered a framework for analysing social relations that historians in particular found attractive. Thus, normal homesteads were defined as functioning with 'primitive communism', or the 'domestic' mode of production, whereas small chiefdoms had a 'lineage' mode of production and large chiefdoms and states operated with a 'tributary' mode. According to one view of this paradigm (Hall 1987), Early Iron Age farmers shared the domestic mode with hunter-gatherers, and the lineage mode only developed at about AD 900 along with the increase in cattle herds.

However, even if we disregard the Central Cattle Pattern, there are still theoretical and empirical reasons why hunter-gatherers and early farmers would not have had the same social relations. For one thing, ownership of small stock and agricultural produce contrasts with egalitarian attitudes towards natural resources. Further, storage of food supplies in above-ground granaries and below-ground pits contrasts with habits of immediate consumption. And thirdly, law courts in settled communities indicate a different way of resolving disputes than that used in societies with temporary camps and loose band affiliations. For these reasons alone, hunter-gatherers and early farmers could not have shared the same mode of production and social relations, whatever they were.

What is more, Hammond-Tooke (1984) invalidated the notion of the lineage mode of production by demonstrating that lineages are not corporate bodies in farming societies in southern Africa. And finally, chiefdoms of whatever size are based in part on the redistribution of tribute. Under scrutiny, then, the modes-of-production framework did not prove useful because it did not fit the data.

Broederstroom is now located on private land and, although visits to the site are possible, they are not easy to make. In any case there is little to see from a tourist's viewpoint. The significance of Broederstroom lies, rather, in its contribution to our understanding of pre-colonial farming societies in southern Africa.

Fig. 9.8   The Broederstroom Early Iron Age site, Hartbeespoort Dam

CHAPTER 10

TSWANA HISTORY IN THE BANKENVELD

Simon Hall

Images of the past

In November 1829, Robert Moffat, on the first of his journeys to the Ndebele state of Mzilikazi, travelled through the Magaliesberg just on the northern edge of the 'Cradle of Humankind'. Here he was astonished to find 'the ruins of innumerable towns, some of amazing extent…[which] exhibited signs of immense labour and perseverance, every fence being composed of stones, averaging five or six feet high…' He continues in his journal to extol and enthuse about the 'style' and the 'taste' of the architecture and some of the 'houses, which escaped the flames of the marauders' and their 'melancholy devastations'. In pensive mood, he sifts through 'these scenes of desolation, casting my thoughts back to the time when [they] teemed with life and revelry…' (Wallis 1945). In a less romantic mode, Andrew Smith in June 1835 was equally impressed and recorded that '[t]he slopes of the hills and knolls were densely covered with ruins of large stone kraals which at the time they were occupied must have contained a great number of inhabitants, though at the time we passed among them not a human being was to be seen' (Lye 1975).

Moffat and Smith had every right to be impressed because even as ruins, these Tswana towns provide a remarkable legacy of a rich and vibrant history. They had a clear sense of this, even though they must have seen only a small number of these Tswana settlements. Archaeological survey shows that they litter the Magaliesberg landscape (see Figs. 10.1 and 10.2). In contrast, Smith was clearly unimpressed with Mzilikazi's settlement and notes that '[t]here was nothing in the appearance of the kraal calculated to impress us with importance of its inmates and, had it not been pointed out to us as the royal residence, we should doubtless have regarded it with indifference…' (Lye 1975). Both travellers witnessed life in the Magaliesberg during a critical cusp in the region's history. Mzilikazi had established the Ndebele state in the 1820s, and in so doing had subdued, incorporated and reorganised most of the Tswana residents into his new political structure; the Tswana chiefdoms and their towns never recovered from the effects of this encounter.

It is a curious fact that the sense of history and teeming life and revelry experienced by Moffat and Smith is all but absent in the present time, despite the continuing and overwhelming presence of these Tswana settlements on the Magaliesberg landscape. Where has this history gone, and why does it not capture our minds today? What exactly happened to these towns and their people? What were the forces that brought them to an end and ushered in a new order? How did these people live and where did they originally come from? How has the history of these large Tswana towns – in which twelve to twenty thousand people lived, and from which powerful rulers organised people, managed large-scale agricultural systems and manoeuvred against competing chiefs for the acquisition of commodities, their production and trade – become mute and all but lost from popular view?

There are several reasons why this history has been subdued. One is a popular perception that nothing much went on in the African past because people were mired in timeless custom. In the colonial imagination, people lived in 'Darkest Africa'. This is one of several enduring images of the African past that in part stem from the narrow view that history can only be written from documentary evidence; consequently, the term 'prehistory' is used for the human past prior to the time when written evidence of this past was produced. This is unfortunate, because it relegates much of the African past to the outer margins of 'real' history. In this curiously static space Africans had 'tribal

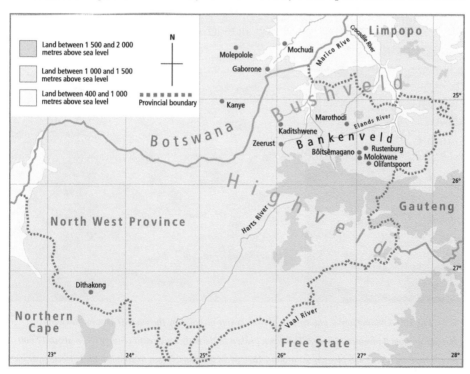

Fig. 10.1 Map of the Magaliesberg showing the location of some of
the Tswana settlements discussed in the text

practice' and lived by superstition in a 'primitive' state, in contrast to Europeans who had 'real' history and lived in a spirit of enlightened and 'progressive' rationality.

A more specific and insidious prejudice is attached to this image – one that was wholeheartedly encouraged and formalised by the apartheid government. This is the belief that people exotic to Africa, often from the Mediterranean world, were the builders and occupiers of African settlements; this substitution of 'outside agents' for the African people who were in fact responsible for these settlements is a way of severing Africans from their own achievements. The monumentality of Great Zimbabwe, the capital of a Shona-speaking state between AD 1290 and 1450, is a classic example. Here the colonial mind de-coupled Shona speakers from their own material history because such achievement was thought to be beyond indigenous ability. The Tswana settlements

Fig. 10.2 Oblique aerial photograph of Molokwane occupied by the western Tswana Bakwena Bamodimosana Bammatau lineage. This town peaked in the early nineteenth century when about 12 000 people lived there and was abandoned around 1826 when Mzilikazi began to establish his Ndebele state in the region. The large linked enclosures in the centre are the cattle kraals of the kgosing and their large size reflects the status and political power of the chief.

that are the focus of this chapter have been treated in a similar way. One example comes from the pen of P-L Breutz, a German ethnologist who was employed in the South African Department of Native Affairs after the Second World War. Over a period of thirty years Breutz worked extensively among Tswana speakers and compiled detailed genealogies and oral histories of most Tswana lineages. Much of this work was published and is of great value to historians, and increasingly so as a parallel source to the archaeology of Tswana speakers over the last three hundred years. Although Breutz gained an intimate knowledge of Tswana history through their oral records, and implicitly established the complexity of Tswana links to this landscape, he persistently rejects the evidence of his life's work, that Tswana speakers were the occupants of the 'stone kraals' that Moffat and Smith observed not long after their destruction (Breutz 1958). Breutz's blind spot continually surfaces in his writing, and so '[s]tone structures are usually of an older culture' he writes (Breutz 1987: 393), that date '[b]efore the time of the principal early Bantu immigrations' and '[t]he decay of the stone builder culture appears to be contemporary with the arrival of certain very early baSotho and maKgalagadi immigrants' (Breutz 1958: 115). He concludes with exceptional irony that '[t]he guess by Archaeologists [that Tswana speakers constructed and lived in these settlements] ignores HISTORY' (Breutz 1987, emphasis in the original), despite the clear connections made by his own informants to these settlements.

Breutz's prejudice, which alienated a significant part of his life's work from its archaeological base, may be amusing in the light of our contemporary political context that seeks an inclusive empathy with the African past. But historical amnesia and prejudice about the past still linger and in some areas continue to flourish. We assume, for example, that the designers of the 'Lost City' hotel, in the heartland of the Tswana world, were not serious about their exotic concoction. Or were they? Perhaps for the popular imagination the historical realities of the African past pale into insignificance in the face of such a fantasy past filled with ancient 'pre-Bantu' civilisations, Dravidian temples and sites that are deemed to predict astronomical events. But such fantasy mocks history, and demeans the present by perpetuating the crude prejudices of the colonial mind. In contrast, a combination of oral histories, early travellers' reports and ethnography makes the link between Tswana speakers and this archaeological evidence unassailable. Archaeology becomes an increasingly important research method as we go further back in time, making it possible to provide an essential outline of early Tswana history. This combination identifies several key historical phases through which Tswana speakers in the Magaliesberg passed, and reveals that there were periods of rapid change, particularly in the eighteenth century, that saw significant developments in the scale of political organisation. The evidence also shows that the Magaliesberg was a significant place for inter-regional interactions, as people were continually drawn to its diverse richness. As historians it is our responsibility to make this history accessible to a wider audience and it is to this history that we now turn.

Early years in the Bankenveld

Archaeological evidence, upon which the earlier phases of farmer history in the Magaliesberg are written, obviously cannot construct a pre-colonial history in the same detail provided by written evidence. We cannot, for example, identify what specific Bantu languages the first farmers in the Magaliesberg spoke. Archaeological work nevertheless clearly identifies the first appearance of these farmers (see Chapter 9). The ability to resolve the identities of second-millennium farmers increases substantially, and much of this period is concerned with ancestral Sotho/Tswana identities. Ancestral Sotho/Tswana speakers did not, however, develop from first-millennium farmers in this region, and we have to look to Limpopo province in the fourteenth century to find out where they came from. In this area small, cattle-centred homesteads with new ceramics styles provide the first evidence for the appearance of ancestral Sotho/Tswana people, and these ceramic styles can be tracked through into the nineteenth and twentieth centuries, and in some areas Sotho/Tswana potters still make them today. Historical linguists also tell us that the closest Bantu languages to Sotho/Tswana are to be found in East Africa, a valuable piece of evidence that indicates that these people did not have their cultural roots in the local Early Iron Age of the Cradle area.

By AD 1500, Tswana-speaking farmers had spread southwards to the southern margins of the Bushveld and westwards into eastern Botswana (Fig. 10.1). This was a steady process driven by a range of 'normal' social, political and environmental factors such as homestead fission and the continuous budding off of independent homesteads. In the Magaliesberg and the ecotonal strip between present-day Pretoria and Zeerust they consolidated their agricultural grip on this highly productive environment (Fig. 10.3). The importance of this period for the political and economic consolidation of ancestral Tswana speakers is underpinned by the creation myth of Matsieng, the 'first' Motswana, who emerged from rock sumps in eastern Botswana. Although mythological in nature, the structure of this story makes historical statements about the ranking of Tswana lineages, and seniority accordingly resides in the Hurutshe of the Zeerust and Madikwe area.

The oral records in their more literal form further lift the veil of archaeological anonymity and reinforce each other in outlining what happened next. As discussed in Chapter 9, the Magaliesberg valley was the southern limit of Early Iron Age settlement in the first millennium AD. The archaeological and oral evidence dating from the

Fig. 10.3 Excavations exposing the remains of an ancestral Tswana house in Madikwe, North West province. This settlement dates to about AD 1600 before stone walls were used to enclose the homestead and enclosures within it, such as cattle kraals.

sixteenth and seventeenth centuries describes and demonstrates that Sotho/Tswana lineages breached this ecological margin and started settling the different grassland habitats of the highveld to the south. By 1650, Tswana speakers had successfully settled these predominantly grassland habitats and had pushed up against the southern limit of summer rainfall agriculture that is roughly defined by the 500 millimetre isohyet. Again, archaeology cannot resolve the specific agencies and motivations that drove this expansion, but we can make some suggestions. One is that this southward move correlates with a short warmer and wetter period within the 'Little Ice Age', an extended period of globally colder conditions that started in the fourteenth century. An agriculturally difficult landscape may have been easier to farm at this time, although in the wider picture of African cattle management, and the variety of tolerances bred into African cereals, it is difficult to imagine that the grassland beyond the bushveld could not have been farmed at any time. Whatever encouraged this expansion, it does perhaps express a different cultural attitude to the landscape compared to that held by Early Iron Age farmers of the first millennium. We must also remember that the grasslands were not devoid of people. Farmers had to renegotiate relationships with hunter-gatherers that ranged between mutually beneficial trade and barter, ritual sanction and more subservient clientship arrangements, particularly among the western Tswana.

The consolidation of Tswana farmers in the Bankenveld, and the extension of Tswana agriculturists to the limits of viable sorghum and millet agriculture to the south, coincide with the increasing substitution of dry stone walling for the pole-and-thatch stockade fences used to construct the central cattle kraals and perhaps the outer boundaries of individual homesteads. This simple raw material substitution is of great benefit for archaeologists, because settlements that include stone in their design are easy to find on the ground and on aerial photographs. Building the basic boundaries of homesteads from stone obviously made good sense in grassland habitats, where wood was a limited resource and fundamental to several key energy needs. Tswana homesteads in the Bankenveld, however, where wood was more plentiful, were also built from stone. In these areas this innovation perhaps has more to do with conserving wood, rather than with the problem of limited supplies. Aerial photographs of the Magaliesberg valley and the hills and ridges of Gauteng provide a remarkable record of these settlements and their high density (see Fig. 10.2). It seems that through the seventeenth and into the eighteenth centuries, the number of farmers on this landscape steadily increased, and the more intensive use of stone underpinned their continuous re-evaluation of sustainability.

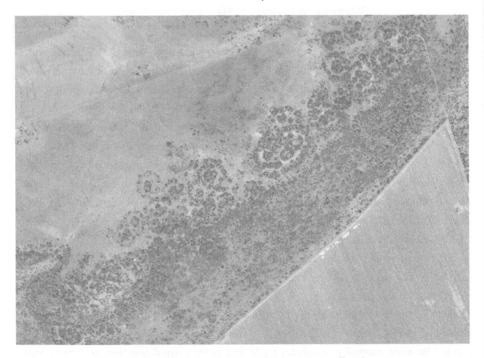

Fig. 10.4 Aerial photograph of a cluster of Tswana stone wall settlements

Political centralisation and the rise of Tswana towns

Up until the early eighteenth century Tswana chiefs exercised relatively low levels of political power. A chief's hold on political status could be tenuous and frequently contested. Succession disputes could have been one of the mechanisms that contributed to the growth in settlement numbers, and continuously worked against the consolidation of power. Although the oral records of the Hurutshe in the Marico area, the Kwena and Fokeng near Rustenburg and the Kgatla further east clearly describe centres of political power, the archaeology tempers over-interpretation of the scale of that power. Historical records show that formal political power was based on the unequal distribution of wealth in cattle, which were central to all social, political and ritual transactions. Wealthy men could transact with bridewealth in cattle for many wives and the legal hold over offspring. Large cattle holdings could also be loaned out in credit arrangements to less wealthy men, and this further increased wealth through interest on these loans. In general cattle-wealthy chiefs had the ability to 'accumulate' people.

Differential wealth in cattle can be archaeologically identified simply by measuring the size of the central cattle enclosures. The archaeological evidence shows that in the early part of the eighteenth century there was little difference in the sizes of cattle enclosures between homesteads, and indicates that there were no great differences in wealth and political power (Fig. 10.4). The number of people living at any one settlement can also be assessed on the basis of the size of single homesteads or the number of homesteads that clustered together. As the eighteenth century progressed, however, it is possible to recognise that homestead clusters increased in size. A number of factors combined to drive this process throughout the eighteenth century, particularly in the second half of this period which the oral records describe as one of increased tension, in which there was cattle raiding and economic rivalry between competing chiefs.

One contributory factor must have been population increase and competition over agricultural space. The Magaliesberg, as already mentioned, was agriculturally rich and within the limits of Tswana farming practice, populations would have risen progressively. It is possible, however, that in the first half of the eighteenth century critical thresholds between available space and agricultural production were reached, in which more coordination was required between individual homesteads in managing agricultural production. The situation may have been exacerbated by the arrival of the BaPô in the Brits area, and other southern Ndebele (the Tswana name, Matebele, for these people has complex origins but more recently is popularly rendered as 'raiders' and 'mercenaries') such as the Tlhako, whose ultimate origins lay in KwaZulu-Natal. Musi is known as a founding chief and Tshwane, his son, settled near present-day

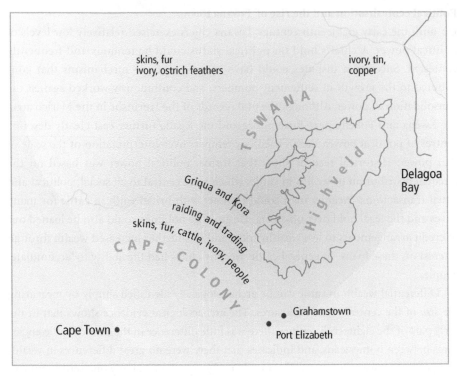

skins, fur
ivory, ostrich feathers

ivory, tin,
copper

T S W A N A

H i g h v e l d

Delagoa
Bay

Griqua and Kora
raiding and trading
skins, fur, cattle, ivory, people

C A P E   C O L O N Y

Grahamstown

Cape Town •

Port Elizabeth

Fig. 10.5 *Map showing the relationship of the Tswana world to the encroaching world
of the Cape frontier and the increasing mercantile interests coming from the east*

Pretoria.[1] Their movement into the northern fringes of the Magaliesberg was perhaps
motivated by the agricultural security of the region. Although these Nguni immigrants
retained their historical identity through oral records, and this continued to be
expressed in the independence of chiefs and their capitals (for example Wolhuters Kop
near Brits), they adapted culturally and became 'Tswana', a process that must have been
underpinned by considerable intermarriage. They spoke Setswana, and built their
settlements and made pottery according to Tswana principles. For these minority immigrants the decision to become Tswana must have been a strategic choice, one based on
rational economic practicalities that facilitated their regional integration. That they were
successful is beyond doubt, for Mogale is remembered for his wealth in cattle and for the
extensive trade that he and his followers were engaged in. Moffat lyrically records the
wealth and abundance described to him by his informants: 'There lived the great chief
of a tribe among his thousands whose cattle were like the dense mist on the mountain
brow,' and 'the numerous ruins through which we daily passed were once very populous,

like locusts…[and] they were exceedingly rich in cattle, grain etc., and trafficked abun-
dantly with the tribes farther in the interior' (Wallis 1945: 9).

The mention of trade introduces one of the central factors in the historical dynam-
ics of the eighteenth and early nineteenth centuries, glimpsed through the oral records.
The records indicate that the encroaching colonial frontiers from the Cape and the
south-east African coast were accompanied by hunger for trade goods from the interior.
Already strong trade relationships, premised on deep time networks, intensified; and
chiefs in the interior competed aggressively for resources, labour and trade routes. Oral
records indicate that fighting and raiding between chieftains was a central characteristic
of the period and aggression revolved around cattle, the prime currency of Tswana
power. Documentary sources that also begin to supplement oral and archaeological
evidence tell of 'a sudden demand for ivory', and those living on the Cape frontier were
increasingly keen, through force and barter, to pass on cattle to the Cape Colony and
obtain ivory, metal, prestige furs, hides and ostrich feathers in exchange (Fig. 10.5). The
records of copper production, for example, indicate that the southern Tswana world was
supplied from the Marico region and that this copper was also traded all the way down
into southern Xhosa communities in the Eastern Cape. The records intimate that
Tswana people were producing goods surplus to their own demands that were traded
regionally, to even out resource gradients, and beyond, moving into the widening maw
of the colonial frontier.

Although the rise of towns and the success of new chiefdoms must have been premised
on winning local conflicts and commanding the power to acquire and control resources,
both for internal consumption and for external trade, the Tswana chiefdoms of the
Rustenburg and Zeerust regions did not only have to contend with each other but also with
Kora and Griqua forces from the south. These people, on the frontier of the Cape Colony,
were mounted and armed, and raided Tswana chiefdoms just as much as they traded with
them. Tswana chiefs were keen to obtain these new instruments of power, not only to com-
pete with the Cape raiders but to gain an advantage over their own Tswana neighbours.

We should perhaps be cautious in reading the oral records as a narrative of continuous and universal conflict and aggression. These are the records of ruling lineages and
are perhaps selectively remembered in ways that bias our perception of the eighteenth
century. Furthermore, oral records are relatively abundant for this period and may falsely
emphasise a change simply because the equivalent oral records are not available for
earlier periods. But the archaeological record provides independent control, and con-
firms the gathering intensity of inter-chiefdom rivalry, because demographic pressure
and trade competition are physically expressed in the emergence of Tswana towns.
These dense and large aggregations of homesteads were flourishing by the end of the

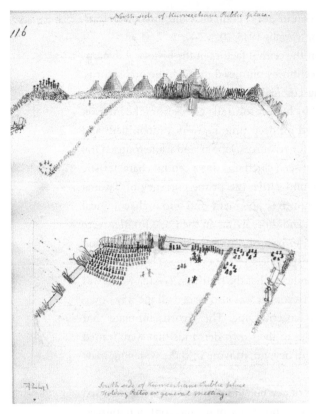

Fig. 10.6 An illustration by John Campbell done in the early 1820s of the central court and assembly area at Kaditshwene, the Hurutshe capital near present-day Zeerust

eighteenth century and were the largest 'urban' centres in southern Africa at this time. Early European travellers confirmed their scale and complexity. Moffat and Smith should not have been surprised by the extent of the Tswana towns they encountered because John Campbell, for example, a member of the London Missionary Society, had previously visited the Hurutshe capital of Kaditshwene in 1820, and left a detailed record of his visit. He estimated a population in the region of twenty thousand people, larger than contemporary Cape Town (Fig. 10.6). In the Rustenburg region, Bakwena Bamodimosana chiefdoms lived at Molokwane and Bôitsêmagano, not 10 kilometres apart, but each with populations possibly in the region of ten to twelve thousand. The Tlokwa capital of Marothodi, occupied between about 1810 and 1827 near the Pilanesberg, provides another example, and other large complexes are found further east both within and on the northern fringes of the Magaliesberg (Fig. 10.7). The rise of these towns represents a complex interplay of increasingly competitive relationships between chiefs, and there is a good correlation between this process and the shift towards town living, with in some cases substantial parts of single chiefdoms gathered at a single place.

The scale of political power expressed in the homesteads (*kgosing*) of the chiefs clearly stands out in these towns. At Molokwane and Marothodi, for example, the cattle kraals of the chiefs are huge compared to commoner homesteads in the same towns, and their power was clearly linked to wealth in cattle (see Figs. 10.2 and 10.7). In comparison, while there are large Tswana homestead aggregations on the Vredefort Dome and Klipriviersberg, for example, the equivalent physical expression of a dominant chief's homestead is not that apparent (see Fig. 10.4). It has been suggested that the shift

Fig. 10.7 Aerial photograph of one of the high-status homesteads in the Tlokwa town of Marothodi, occupied between about 1810 and 1827

towards aggregation expresses a need for defence, and the position of some aggregations and towns on more easily defended hills reinforces this. Other towns such as Molokwane and Marothodi, however, were not located on hills, which indicates that these chiefdoms were politically strong and confident about their regional power. The oral records add a corroborating strand to this suggestion. At Molokwane, for example, Kgasoane,[2] who was born in about 1740 and became chief in about 1780, was known as a powerful and successful chief in the region. In spite of considerable regional conflict and succession disputes he was able to maintain the position of Molokwane in the landscape without resorting to the defensive positions of hills. The Tlokwa capital of Marothodi, occupied between about 1810 and 1827, provides another example. Marothodi was built on the flats between the Pilwe and Matlapeng Hills near the Pilanesberg, in a patently undefendable position. In contrast, the earlier Batlokwa settlements of Maruping and Mankwe had been defensively located on the southern side of Pilwe Hill, some 8 kilometres to the east of Marothodi. These earlier settlements date to a festering conflict between about 1780 and 1810 between the Batlokwa and the Bafokeng to the south. With the aid of Kgatla allies, the Tlokwa eventually overcame the Bafokeng. The open location of Marothodi (see fig. 10.7) suggests that after this conflict they were politically powerful and economically secure in the region.

Although the context within which Tswana towns developed was one of competitive and hostile relationships over control of space, people, resources and trade, everyday routines continued, and the bottom line of town life was that communities had to feed themselves. The ability to farm successfully for dense populations may have been helped by increased rainfall during the eighteenth century, but droughts early in the nineteenth century may have added additional stress. If maize, introduced to the south-east African coast by the Portuguese perhaps as early as the sixteenth century, had made a significant contribution to the inhabitants' cereal diet by the nineteenth century, the effects of drought on conflict would have been intensified because maize requires higher rainfall than sorghum and millet. Even if maize was not under cultivation in the Magaliesberg, the ripple effects of drought on maize agriculture in the higher rainfall areas to the east may have been felt further afield. A re-examination of Figures 10.2 and 10.7 reminds us that the scale of agricultural production required to feed the large populations in towns such as Molokwane, Bôitsêmagano and Marothodi must have been considerable. Although chiefs ritually coordinated the agricultural cycle, individual homesteads were responsible for their own needs. As mentioned earlier, the Rustenburg area was agriculturally rich and it is doubtful whether women moved for several months to fields that were some distance from the town, as is the historic practice among Tswana speakers in the more arid areas further to the west. The extensive and rich soils in the immediate vicinity of these towns suggests that fields were close at hand. Given the premium placed on cattle, it is probable that these were also managed locally, rather than at distant cattle posts. Additionally, huge pressure must have been placed on other essential resources such as wood for domestic needs and basic household construction. The use of stone to build homestead boundaries and cattle enclosures obviously saved wood and reminds us that the initial choice to use stone, and its gathering intensity through the seventeenth and eighteenth centuries, was underpinned by a clear understanding of sustainability. The character of this period is basically one of exploitation of resources and management of a more competitive market. It also cautions against the notion that these societies lived in idyllic harmony with 'nature'.

While the agricultural potential of the Magaliesberg and surrounds was relatively even throughout, the same does not apply to other resources. Archaeological work is just beginning to hint at the details of local commodity production, local specialisations and possible regional exchanges. Iron, for example, was obviously a critical commodity for agricultural work, and the manufacture of tools and weapons and the consumption at a town the size of Molokwane was considerable. It is intriguing, however, that archaeological work there has produced no evidence for iron production. If the residents of the town were not responsible for producing their own iron, who supplied them? These are not easy questions to answer, but a combination of the archaeology and oral records provides some contexts within which to develop ideas. We have already mentioned that Kgasoane was a powerful chief during the period when Molokwane achieved its maximum size in the late eighteenth and early nineteenth centuries, and such power, as

suggested by the oral records, must have been an important ingredient in negotiating regional trade and even commanding tribute. Archaeologically, a town such as Marothodi provides a significant contrast with Molokwane because the scale of copper and iron production there suggests surplus metal production for regional trade (Figs. 10.8 and 10.9). Although trade links to Molokwane still need to be demonstrated, the uneven distribution of vital resources provided opportunities for specialisation in commodity production. Hide production may have been another such contributor to regional trade. Early European travellers describe and depict large-scale hunts in which considerable labour was mobilised to drive game into extensive pit-fall traps (Fig. 10.10). Large quantities of meat and hide would have resulted, and because communal hunting was controlled by chiefs and hide processing was male work, surplus production and trade must also have been controlled by men. At all Tswana settlements dating to this period that have been archaeologically investigated, tools for hide preparation have been

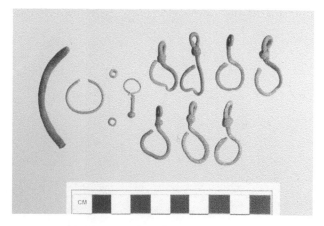

Fig. 10.8 A cluster of copper refining furnaces at Marothodi

Fig. 10.9 Bronze (an alloy of copper and tin) and copper earrings and copper beads and bangle pieces from Marothodi

found, but at Olifantspoort, a large Bakwena settlement some 30 kilometres east of Molokwane (see Fig. 10.1), evidence of what may be a specialised hide production area has been found. Whatever aspect of the economy we consider, the scale and extent of production during the town phase was significant, and the physical expression of political power around which production hinged is also evident. When we compare these late eighteenth- and early nineteenth-century Tswana towns with the more dispersed earlier settlements, the change in economic scale is clearly evident.

Mzilikazi and the Ndebele state

We have emphasised that the rise of Tswana towns was a response to a number of factors, one being the encroaching colonial frontiers. Consequently, although the development of these towns was a local Tswana response, the historical circumstances that contributed to their birth were common throughout southern Africa. The rise of Shaka and the Zulu state in what is today the province of KwaZulu-Natal is an example of another distinctive local response to the same general pressures and opportunities. The history of Tswana towns in the Bankenveld cannot be separated from a much wider context, and this applies even more so to the events of the early part of the nineteenth century.

As noted earlier, Moffat, Smith and other early European travellers in the

Magaliesberg and Bankenveld from the 1830s recorded the extensive ruins of Tswana towns. In concluding this discussion, we need to return to a consideration of why these towns were abandoned. The oral records indicate that by the early nineteenth century, the scale of some conflicts between Tswana chiefdoms had contributed to their own destruction. Already weakened by internal disputes and famine, Tswana chiefdoms could not resist the military organisation of Mzilikazi and his Ndebele, who arrived in the Rustenburg area in 1827. Moffat records that the Tswana of the Magaliesberg 'had become effeminate by peace and plenty', but in some quarters still boasted that 'their numbers would awe their enemies' (Wallis 1945: 9). Mzilikazi, however, was a strong-willed Ndebele chief who had left Natal after a dispute with Shaka, and who sought an area to establish his own independence. A common image of his rapid domination of the Tswana in the Rustenburg area is one of brutal and bloodthirsty conquest. This is not entirely fair, because viewed from another perspective, Mzilikazi brought order to a Tswana world that was struggling under its own strife. The organisation of the Ndebele state offered protection from the increasing and ruthless encroachment of Griqua and

Fig. 10.10 An illustration by the traveller AA Anderson which captures a coordinated pit-fall hunt, in which people drive animals towards an extensive field of pit-fall traps

Fig. 10.11 Photograph of 'huts' that form part of a larger underground settlement called Lepalong, which was occupied periodically by Tswana speakers during the difaqane (the 'time of troubles') between the mid-1820s and late 1830s.

Kora raiders from the south of the Magaliesberg, and stability in economic life. Under the cultural domination of the Ndebele, many Tswana became Ndebele, while other chiefdoms, such as the Hurutshe at Kaditshwene, struggled on with nominal independence until the early 1830s. Belonging within the Ndebele state meant eschewing prior identities, and Tswana people learnt the Ndebele language, dressed in the Ndebele way and lived in their quite distinctive settlements.

Another option for Tswana chiefs was to relocate outside the borders of the Ndebele state. The Bakwena Modimosana, for example, abandoned Molokwane after 1827. Kgasoane was old and frail at this time and died while retreating southwards out of the Magaliesberg and onto the highveld. The new chief, Maseloane,[3] was one of several leaders, including the Tlokwa from Marothodi, who in a much reduced state periodically took refuge in an extensive cavern system called Lepalong near present-day Potchefstroom (Fig. 10.11). This refuge does reflect the dire straits of Tswana communities outside the protective power of the Ndebele state, as well as the more widespread uncertainty during this 'time of troubles' (*difaqane*).

Mzilikazi's power in the Rustenburg region was constantly under threat from all quarters, particularly the colonial and Boer trekker frontier to the south. He progressively moved the nucleus of the state further west, and eventually relocated to south-western Zimbabwe. Tswana communities did re-establish themselves to a certain degree, but hopes of regaining former power were progressively thwarted by the arrival and establishment of the Voortrekkers, who by the 1850s came to dominate the Bankenveld. With little chance of recovering their independence, certain chiefs such as Pilane of the Kgatla and Gaborone of the Tlokwa moved west into Bechuanaland. Towns such as Mochudi in Botswana represent continuity with the tradition of Tswana towns started in the eighteenth century.

An image of Tswana history

This chapter has briefly highlighted some of the more prominent episodes and themes of Tswana history in the Bankenveld. Constructing this history depends on the combined use of several sources. Archaeological evidence is at the core of this review. This is animated by oral records and some early traveller diaries, and the ethnography of Tswana speakers provides critical information on the cultural values expressed in the archaeology. This history comes increasingly into focus when the oral records provide detail on key players – the names of chiefs, key conflicts and central alliances. The oral testimonies put a historical 'face' to the development of Tswana towns in the eighteenth century, and in combination with archaeology, belie an image of a static past. Although we do not have the same detailed records for earlier periods, this does not make them any less part of the historical process in the region. The rise of Tswana towns is a history of considerable scale which cannot be reduced to an image of communities living in isolation; they are not, as Breutz would have it, the remains of a 'pre-Bantu' past. These communities were intensively connected to others at local, regional and sub-continental scales. To glimpse this connectivity is critical, because it again subverts the notion that Africans were captive to their own tribal condition. On the contrary, what this history emphasises is that people change their cultural responses in accordance with historical conditions, and the Tswana in the Bankenveld are no exception.

CHAPTER 11

THE EARLY BOER REPUBLICS:
CHANGING POLITICAL FORCES IN THE 'CRADLE OF HUMANKIND', 1830s TO 1890s

Jane Carruthers

Two forces drove the frontier of white expansion and settlement forward into the interior of southern Africa in the late eighteenth and early nineteenth centuries. The first was the attraction for pastoralists of grazing lands for their herds and for hunters of ivory and skins; the second a desire on the part of the colonial government at the Cape to have reliable scientific knowledge about the people and natural resources of the region.

Trekboers and explorers

The trekboer movement had begun many decades earlier, as a number of factors both pushed the less wealthy farmers out of the confines of the Cape Colony and propelled them further northwards and eastwards. Dependent on a life of semi-nomadism or transhumance so as to be able to move at liberty to better grassland when it became seasonally available, small communities of Dutch/Afrikaans-speaking pastoralists had sparsely populated the interior by the end of the 1700s. Because of the increasing distances between them and colonial urban centres, the trekboers came to rely almost completely on the natural resources of the area for their physical sustenance. They made extensive use of animal hides for ropes, blankets and even for clothing, and used local materials for shelter and other basic requirements, thus limiting their expensive imports to essentials such as guns, gunpowder, coffee, tobacco and sugar (see van der Merwe 1938, 1945; Guelke 1989: 90–91). Etherington explains that while many trekboers might have aspired to 'stay put and make a modest living', the land could not support a dense population and groups therefore remained small and relatively mobile (Etherington 2001: 48–49).

The second force that motivated exploration into the interior was the need for scientific information about the people and resources – especially wildlife and the rumoured mineral wealth – of the sub-continent. Once the Cape came into British hands after 1795, exploration into the 'Far Interior' was facilitated by the new government with its objective of deliberate colonial settlement and the maximisation of economic opportunity and mercantile penetration; this was a different approach from that of its predecessor, the Dutch East India Company. Governor Francis Dundas, for example, despatched an expedition into the north-west so as to stimulate the inland trade in livestock, because drought and wars in the eastern regions of the Cape had led to a shortage of cattle in the colony. On this expedition, which lasted from October 1801 to April 1802, Pieter Truter and Dr William Somerville and others, accompanied by the painter Samuel Daniell, journeyed into Tswana country, recording local languages and customs, noting sources of water and, most significantly, becoming cognisant of the great abundance of wildlife that existed beyond the colonial borders (Bradlow & Bradlow 1979; Parsons 1995: 345). John Barrow, a British bureaucrat and early South African cartographer, was another official who travelled extensively in search of a better knowledge of the geography and resources of the sub-continent (for details of his travels see Barrow 1801–1804).

DRIVING IN AN ELAND

Fig. 11.1 'Driving in an eland'. Illustration by William Cornwallis Harris, mid-1830s

In the second half of the 1700s, the trekboer movement also threw off ahead of it a number of dispossessed Khoi/San inhabitants of the regions through which it passed, along with a number of sometimes mixed-race individual traders, adventurers and renegades who penetrated deep into the interior, using firearms and force to forge new societies. !Kora intruders drove further and further north from 1760 until eventually reaching as far as the Nkwaketse. They were joined a decade later by *oorlam* coloured hunter-traders (Parsons 1995: 344–345). One of the most notorious of the European frontier bandits was Jan Bloem, a German deserter and criminal, who attracted a large number of mixed-race followers – including escaped slaves, dispossessed Khoisan, criminals and misfits – on the colony's northern Cape frontier. Becoming 'a powerfully disruptive force', Bloem and his men raided !Kora groups in the northern Cape and even extended their operations against the Sotho–Tswana to the north, raiding their cattle (Penn 2005:

198–199). In the northern Transvaal area and along the Limpopo River valley, Coenraad de Buys made his mark. Having settled intermittently among the Xhosa, Thlaping and Rolong, he and his large number of sons and their followers were 'active participant[s] in the rising violence' (Parsons 1995: 345–348).

As pressure was exerted on the African communities north of the Vaal River and as the population increased, the opportunities for conflict escalated. Violence was exacerbated by access to guns and the use of horses. No one group was able successfully to annihilate all opposition, as the various factions were generally evenly balanced and life was continually insecure. Nevertheless many groups, such as the Hurutshe at Kaditshwene, regrouped and retreated to more defensible and hilly locations (Boeyens 2003).

In 1823 the final source of dislocation arrived in the area – Mzilikazi and his Ndebele kingdom, who had taken flight from conflict in the Zululand (KwaZulu) region. Mzilikazi first centred his chiefdom on the middle reaches of the Vaal River in the vicinity of what later became Heidelberg. There he found himself repeatedly raided by the Taung leader Moletsane, and by Griqua and !Kora bandits. Continuous harassment persuaded him to move to the north and the west, where he sited his capital on the northern side of the Magaliesberg range in 1837. A further attack by Jan Bloem and the Taung chiefdom in 1828 convinced him of the need to create a *cordon sanitaire* of depopulated land east and south of previously resident chiefdoms such as the BaPô, the BaFokeng and the BaKwena ba ka Makopa who were forced to relocate and pay tribute. Mzilikazi did not then simply devastate and depopulate the land, as was claimed in most later accounts, even though he cut a swathe of destruction through the area. In the early 1830s, after a renewed attack by the Zulu, he relocated even further north-west in what is today the Marico district (Simpson 1986: 82–84).

A trading frontier opened, too, and by the late 1820s adventurous entrepreneurs such as Robert Scoon and William McLuckie, both farm labourers turned hunter-traders, had explored well beyond the Gariep River and into what was to become the Transvaal. They had encountered Mzilikazi, who had settled with his followers in the highveld and bushveld during the period of the *difaqane* (beginning around 1820, when many interior societies collided with one another, causing a decade or more of intense disruption). Scoon and McLuckie took the missionary Dr Robert Moffat to meet Mzilikazi in 1829. At the same time, trader David Hume was taking the Reverend James Archbell in the same direction. In 1834 Hume joined a famous expedition into the region north of the Vaal River that was led by Dr (later Sir) Andrew Smith. Smith's account of the interior is one of the first formal and comprehensive records about the area located within the region of what is now known as the 'Cradle of Humankind'

*Fig. 11.2 'Wagons on Market Square, Grahamstown' (1850). Thomas Baines's
painting illustrates the enormous extent of the ivory trade.*

(Smith 1849; Lye 1975. See also Brooke Simons 1998.). This work, together with Smith's
zoological descriptions and illustrations (by artists including Charles Bell and George
Ford), remains one of the most important sources of information about southern Africa
at that time. Smith, a surgeon and later director-general of the British army's medical
department during the Crimean War, had started the South African Museum – the
country's first – in 1825.

The information gathered by hunters, traders, missionaries and emerging
communities such as the Griqua gave a picture of an interior burgeoning with attractive
natural resources. This knowledge encouraged the formation of the South African
Literary and Scientific Institution, which sponsored the Association for Exploring
Central Africa with Smith as its leader. Evidently an extremely intelligent man, Smith
was employed by the colonial government to gather ethnographical knowledge about
local people, to study their culture and lifestyles, to assess their responses to colonial
government policies and to collect specimens of natural history. His expedition was
well equipped and it was tasked with collecting scientific information (particularly
concerning natural history and geography), befriending local chiefs, collecting intelligence about African people, and exploring opportunities for commerce and trade

*Fig. 11.3 'The Outspan'. Painting by Charles Bell of Dr Andrew Smith's
expedition of 1834–1836 into the interior*

beyond the confines of the Cape Colony. In the course of the expedition the Smith
party's encounters with Tswana and other chiefs were generally friendly and John
Burrow, a young man in the group, recorded his impressions in a letter to his parents:
'We have found the Natives civil, hospitable and kind and have met with more assistance from them than we could have got for nothing had we travelled as long in the
Land 'o Cakes [Scotland]…' (quoted in Brooke Simons 1998: 35). Finding rhinoceros
and many other big game species, their horses surviving horse sickness, and observing
the cultures and politics of indigenous people, Smith's expedition was on trek for more
than a year. Having reached their goal, the tropic of Capricorn, by January 1836 the
party had returned to Cape Town. Smith's expedition was of great consequence in collecting an enormous quantity of new zoological and topographical material. Smith,
who had met and admired Charles Darwin, was unusual in retaining type specimens in
a colonial museum – the South African Museum in Cape Town – rather than sending
them to London, as was the more usual approach at that time.

SHOOTING THE HIPPOPOTAMUS.

Fig. 11.4 'Shooting the hippopotamus'. Illustration by
William Cornwallis Harris, mid-1830s

Smith had planned to proceed eastward on the southern side of the Magaliesberg (first referred to as the 'Cashan Mountains' after Kgaswane, a local chief of the Tswana BaKwena ba ka Mmatau) in order to find the source of the Vaal River. Although his Ndebele guides refused to take him in that direction, he managed to get close to the present site of the Hartbeespoort Dam in July 1835. While in the region of the 'Cradle of Humankind', Smith and his men hunted hippopotamus and a good deal of other wildlife. One member of the party, Andrew Geddes Bain, even collected specimens to export to the United States of America.

Glowing descriptions of plentiful wildlife soon brought recreational sport hunters into the interior, and many of the early writings on this theme are iconic in southern African literature. Of these, the books written and illustrated by William Cornwallis Harris are among the best known. *Portraits of the Game and Wild Animals of Southern Africa* and *The Wild Sports of Southern Africa*, Harris's writings about his hunting trip of 1836, had an enormous impact in Britain and were best-sellers. In the Moot, the valley

Fig. 11.5 'Bechuana attacking a rhinoceros'. Painting by Charles Bell, who accompanied
Andrew Smith on his expedition into the interior of southern Africa in 1834–1836

Fig. 11.6 'Hunting the wild elephant'. Illustration by William Cornwallis Harris, mid-1830s

between the Magaliesberg and the Witwatersberg, close to the Crocodile (Oori) River, Harris encountered what he called a 'fairy land of sport'. He gloried in a landscape literally covered with hundreds of elephants and he delighted in their slaughter. Harris was a naturalist as well as a hunter and he identified every wild animal by its correct zoological name, describing its habits and morphology carefully. In this 'menagerie', Harris was thrilled to discover a large mammal that was new to Western science, *Hippotragus niger*, the sable antelope, a great zoological prize known for many years thereafter as the 'Harris buck' (Harris 1840, 1852).

The threat of Voortrekker colonisation

While out hunting and observing big game, Harris also saw many ruins of Tswana cattle kraals, remarking that '[t]hese crumbling memorials now afford evidence of the extent to which this lovely spot was populated before the devastating wars of Moselekatse laid it waste, and indicate also a refinement in the art of building that I had not met with before'(Harris 1852: 158). Mzilikazi, who at that time controlled a vast area of the central and western Transvaal, was suspicious about intruders like Harris, and although he gave the visitor permission to hunt for a month, Mzilikazi had him followed throughout the course of his journey (Harris 1852: 205). Mzilikazi's suspicion of whites was heightened by the growing number of Voortrekkers who entered his territory in the mid-1830s. They did not come by way of the usual Western 'Missionary Road' (nor did they have the approval of the missionaries), but via the 'forbidden' route, directly across the Vaal River from the south, the same direction from which Mzilikazi's Zulu and Griqua enemies had come. Harris's book gives a graphic contemporary description of the Great Trek, that exodus of Boers from the Eastern Cape who emigrated because of land hunger, war-weariness and dissatisfaction with British rule. While writing admiringly of the Voortrekkers' pioneering

*Fig. 11.7 'Aigocerus niger' [Hippotragus niger],
The sable antelope. Illustration by William Cornwallis Harris, mid-1830s, the
first person to describe this antelope for science*

Fig. 11.8 'Hunting at Meritsane'. Illustration by William Cornwallis Harris, mid-1830s

spirit, Harris prophesied that this intrusion into African-held land had 'kindled a flame in the interior which can be only quenched with blood' (Harris 1852: 286–300).

At that time, the emigrating Boers were not traders, trophy hunters, naturalists or ethnographers, but aggressive colonists seeking suitable areas in which they might settle freely, far away from the British government but close to abundant sources of African labour. In their search for land, labour and the profits of the hunt, the Voortrekkers fought many battles against African communities. In 1836, with the help of their BaPô allies, the Boers routed Mzilikazi at Vegkop, and the Ndebele chief and most of his followers retreated into what is now Zimbabwe. When French naturalist Adulphe Delegorgue passed by Vegkop some years later, the terrain was still littered with the bones of the fallen (Delegorgue 1997: 163).

The values that the Boer settlers in the Transvaal brought with them were mercenary and practical: it was not part of their world view to expand knowledge about the sub-continent. In order to survive they were obliged to live off the natural environment, and the mainstay of the early Transvaal economy was an export trade in ivory and hides. Without the wildlife trade and the concomitant destruction of hundreds of thousands of animals, the Voortrekkers would not have been able to exist as independent polities. Further north, near the Soutpansberg, a hunting frontier developed in which Africans and Boers were partners; the Africans were armed and willing to hunt on foot in areas that were death to horses, for in much of the Transvaal at that time horse sickness, nagana and malaria were endemic diseases (for details see Carruthers 1995 and Wagner 1983).

Writing of that time, AA Anderson described 'a country full of large game – elephants, rhinoceros, and giraffe browsed on the banks of the Vaal, down to the Orange River' (Anderson 1974: 268). There were so many wild animals close to the Vaal River that there was a dearth of grazing for Delegorgue's oxen in the early 1840s (Delegorgue 1997: 64). Later in that decade Chapman discussed the economy of Mooi River Dorp (later called Potchefstroom) in terms of its cattle and ivory trade, upon which profits of 75 to 150 per cent could be made, although expenses were very high. Game could be shot without leaving the town, and Chapman was able to keep a menagerie (Chapman 1971: 18). But sport and market hunting by many thousands of armed African and European hunters meant that there was a frenzy of killing, and within a mere twenty years the immense herds seen by Harris – which he had compared in number to swarms of locusts – had gone. Thomas Baines was only one of many who came to South Africa fifteen years after Harris to be bitterly disappointed at the lack of opportunities for sport in the southern and central Transvaal (Kennedy 1964: 54).

Having left British governance behind in the Cape Colony, the Voortrekkers set about creating their own independent community outside of the colonial sphere of influence. The process was slow and riven by factionalism, inexperience and tremendous resistance from African communities. By 1838 a permanent toehold had been secured north of the Vaal River, at Potchefstroom. Just a few years later the 'emigrant community' had spread to the north-east, into the more tropical bushveld to the north of the Cashan Mountains. Called by then the 'Magaliesberg' (after a local BaPô chief named Mogale), rather than Cashan, it developed quickly as an outpost of the Potchefstroom (Mooi River) Republic. The attraction of the Magaliesberg was its perennial rivers and well-watered kloofs, its wooded southern slopes, the fertile soils of the Mooi valley between the mountains and the warm bushveld climate on the north of the range. It was also the ideal place from which to foray further north to raid wildlife and slaves. Prominent Boer families who settled in this district were the Krugers – Gerrit, Casper and his son Paul (later president of the Transvaal Republic); the Potgieters – Hendrik and Hermanus; and the Pretoriuses – Marthinus Wessel (son of the Voortrekker leader Andries Pretorius and later president of both the Transvaal and the

Orange Free State), Bart and Piet (the latter two were the owners of the farm 'Broederstroom'). In many ways it was the heart of the early Transvaal.

The intense rivalry among Voortrekker warlords is the stuff of high drama – as well as of civil war (Etherington 2001: 294–295). Hendrik Potgieter, who had led the successful campaign against Mzilikazi and thus claimed the right of the Boers to occupy the Ndebele kingdom, hated Andries Pretorius, the man who had taken the Voortrekkers into Natal and vanquished the Zulu at Blood River. Potgieter, whose policy was to gain independence by moving well beyond the zone of British jurisdiction (the twenty-fifth parallel) and thus avoiding any dealings with Britain, left the Magaliesberg in 1845 to found another republic at Ohrigstad further north. He appointed Gerrit Kruger, owner of the farm 'Hekpoort' (now a small village just beyond the official boundary of the 'Cradle of Humankind'), as commandant of the Mooi River (Potchefstroom) and Magaliesberg communities in his absence (see Strydom 1955: 68; Potgieter 1958: 50). A church gathering on 'Hekpoort' in 1848 gives an indication of the population at that time, for it attracted about 1 000 people and 209 wagons (Potgieter 1958: 51). Potgieter's rival, Andries Pretorius, whose policy was to maintain Voortrekker autonomy by dealing directly with Britain, either diplomatically or militarily, had been unable to come to a satisfactory agreement with Britain over Natal, which had been annexed in 1843. Seeing no prospect for Boer independence within a British-dominated Natal, Pretorius and a large number of followers had trekked from that colony to the Magaliesberg district in 1848. There he staked out his claim to the farm 'Grootplaas', a property at the confluence of the Crocodile and Magalies Rivers, now partly flooded by the Hartbeespoort Dam.

In order to bring the issue of Voortrekker autonomy to a head, Pretorius chose to confront Britain and in August 1848 he led a commando back across the Vaal River to this end. At the battle of Boomplaats, he and his followers were soundly defeated by a British force under Sir Harry Smith and the area between the Gariep and the Vaal Rivers was annexed to Britain as the Orange River Sovereignty. Clearly, the confrontation policy was not successful. Nor, however, was isolation beyond British jurisdiction, for Potgieter and his followers had been forced to abandon Ohrigstad because of malaria and because the low-lying country to the east and north-east was uninhabitable by whites at that time as a result of human and livestock diseases. There was no doubt that presenting a united front would be more productive than vicious squabbling, and in order to define what Voortrekker policy *vis-à-vis* Britain should be, a number of public meetings were held to try to bring the Potgieter and Pretorius factions together. One of these crucial meetings was held at 'Hekpoort', on the doorstep of the Cradle, on 9 February 1849. There it was decided to accept the authority of a single governing body,

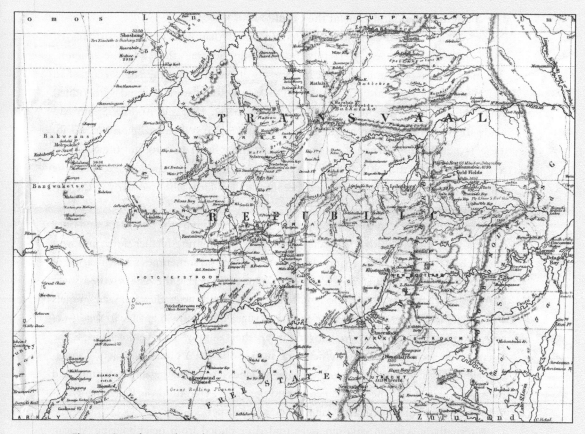

Fig. 11.9 Part of a map drawn by Thomas Baines and published in 1877. It shows the routes taken by travellers and hunters through the Transvaal, traversing the 'Cradle of Humankind'.

and in May 1849 the first united Volksraad (Citizens' Council) sat at Derdepoort, east of present-day Pretoria where the highway to Warmbaths (now renamed Bela Bela) cuts through the mountains. In 1852 Pretorius was given the authority to meet with British representatives on the Sand River, and he negotiated permanent independence for the Boers under the terms of a formal convention known as the Sand River Convention. Thus was made possible a single and united government for the Zuid-Afrikaansche Republiek (ZAR) north of the Vaal River, although it took a good deal of further negotiating to get the Potgieter group to agree. Only after 1860 could the Transvaal be regarded as a nominally unified independent republic under a single government. Up until then and even afterwards, not only the ZAR but even its individual regions remained weak, divided and in an unstable state of friction and interdependence with their African neighbours.

The changing social landscape of the Cradle

Precise details of how the relations between Boers and Africans in the Cradle area played themselves out are unclear. Certainly the whole district was affected by the *difaqane*, and many Tswana communities became vassals of Mzilikazi. In the 1830s Harris had commented that the southern slopes of the Magaliesberg were 'covered with ruined kraals' of people driven out by Mzilikazi (Harris 1852: 158) and Andrew Smith had made the same observation a little earlier (Lye 1975: 83). When Mzilikazi fled north, these Tswana groups (many of which had assisted the Boers to drive out the Ndebele) returned to their land only to find Boers already firmly in occupation. Probably the majority of Africans in the Cradle area became landless farm labourers early in the history of the South African Republic.

Most of the large game on the Magaliesberg was shot out by the late 1840s, and Boer hunters from the Mooi River/Magaliesberg area had to venture further north and west in search of new herds. This required the acquiescence and sometimes some measure of support from westerly chiefdoms like the BaKwena of Sechele. Other Boers settled down to more agricultural pursuits, and Paul Kruger describes how eradicating wild animals was crucial to promoting settled agriculture (Kruger 1902: 18). To engage in this, however, required labour, which was invariably in short supply. To meet such needs the Boers made heavy demands on chiefdoms such as the BaPô under Mogale. Although Mogale had been a good friend to the Boers, his lands along the Magaliesberg range and the river were not returned to his people and the BaPô were put to work digging irrigation furrows and otherwise assisting the Boers to prosper.

Alternatively, African children were traded or raided from neighbouring African chiefdoms. A mainly younger generation of Boers who constituted a kind of brigand-hunter class provoked wars and organised raids on neighbouring African societies, carrying off children whom they apprenticed or semi-enslaved at Mooi River and elsewhere. This led to a near-constant state of imminent hostilities or war. As the hunter-trader Chapman recorded in his journal of the Mooi River/Magaliesberg community in the early 1850s, '[t]he Boers do not behave well to the poor tribes who were independent before the Boers intruded on their domain, in consequence of which they have to continually struggle with one tribe or other in war…they buy the children of some natives, and in war take away as many as they can' (Chapman 1971 Volume 1: 20).

Chapman went on to describe deep rifts within the Mooi River community. 'The young Boers,' he observed, were 'willing to make war for a trifling thing…in anticipation of the booty which has always been divided among them and likewise the native children' (1971 Vol.1: 21). These 'drunkards and riff raff' he contrasted to the 'respectable' faction of settled farmers personified by Jan Viljoen. According to Chapman, Viljoen on one occasion in 1851 accused the leader of a rival faction, Scholz, of being 'the sole cause of losses to the boers and the community of tribes' and warned Pretorius in Pretoria that 'all the tribes would soon join Sechele if Scholz's plans went ahead [with] his unjust war' (1971 Vol.1: 81–88).

Anderson described the situation as follows: 'If they were an industrious and well-disposed people, and cultivated their lands in a proper way, the Transvaal would, and ought to be, the most prosperous and well-to-do country in South Africa...But no, they would sooner expend their energies in fighting the native tribes and stealing their cattle, because it pays them better, than devote their time to peaceful pursuits. From the time the Boers have held the Transvaal they have pursued this policy...' (Anderson 1974: 265). Thomas Baines heard from a resident of the Magaliesberg area that the reason the Boers so feared the British was that it was only after the Voortrekkers had 'cleansed' Natal of Africans that the British had taken it from them and his inform-ant believed that the same was going to happen in the Transvaal, because it too had been 'cleansed' (Wallis 1946: 620–621). Subjugation of Africans often took the form of raids and retalia-tion. For example, in August 1858 Paul Kruger led a raid against a Tswana group accused of murdering two Boer men. After defeating the Tswana, the Boers beheaded the leader, 'Gasibone', and sent the severed head to the superior chief as a warning (Nathan 1944: 61–62).

When Boers first settled north of the Vaal River, race rela-tions seem to have been fluid, and in 1849 Chapman said that one of the Heemraden (elected local councillors) of Potchefstroom named R du Plooy was 'a half-black', who sometimes acted as the landdrost (magistrate) when that official was absent. But within the boundaries of the republic, suspicion bedevilled relations between African subjects and Boer masters. In 1847 an informer told the Boers that Mogale had a supply of guns hidden in a nearby cave, but by the time the Boers arrived, the firearms had been sent to chief Mokopane, one of the Boers' most significant foes to the north, where armed conflict and inter-racial warfare were endemic. Refusing to appear before Gert Kruger for questioning, Mogale and many of his followers fled for safety to Moshoeshoe, the BaSotho chief on the Caledon River on the Lesotho border. They remained there for about fifteen years, and in 1855, in

Fig. 11.10 A statue by Adam Madebe of Mogale, the nineteenth-century chief of the BaPô after whom the Magaliesberg is named.

193

Mogale's absence, Kruger held a sale at 'Hekpoort' of the almost one thousand cattle the Tswana chief had left behind. The following year Kruger wrote to Marthinus Wessel Pretorius about negotiations to allow Mogale's elderly mother to visit her son at Moshoeshoe's settlement, providing she try to persuade him to return to the Transvaal. According to the oral traditions recorded by government ethnologist P-L Breutz in the 1950s, Pretorius had promised that if Mogale returned, all his possessions would be restored to him. Apparently, however, Mogale could not read the letter and thus did not respond. Mogale was in fact anxious to return to his ancestral home and both visited Gert Kruger and sent a placatory gift of four horses to him. But – for reasons unknown, perhaps out of fear of reprisals – Mogale remained only one night in the cattle kraal at 'Hekpoort' before returning to Moshoeshoe for another three years.

Pretorius again tried to tempt Mogale to return with a promise of land in the Heidelberg area or near Olifantsnek (the far western end of the Magaliesberg), but Mogale insisted on residing where his ancestors were buried and in due course, around 1862, purchased the farm 'Boschfontein' (near Wolhuterskop, on the northern side of the Magaliesberg) from a Mr Orsmond for 499 head of cattle (for details see V Carruthers 1990: 264–265; Breutz 1953: 111–112, 179–185; Transvaal Archives TA SS8 893/55, TAB SS0 R1008/56). Mogale's purchase of this land was to have immense symbolic and practical significance. Throughout the late nineteenth century African chiefdoms across much of the Transvaal, but especially in this general region, repurchased land on a very substantial scale. In many instances they taxed earnings of whole groups of migrant labourers working on the mines to generate the necessary funds. Generally they repurchased the land using missionaries as intermediaries. Mogale's successor, Frederick Morvantoa Mogale, for example, purchased several more farms just west of Wolhuterskop through the Hermannsburg Mission Society (Breutz 1953: 182–184). Such purchases continued into the early twentieth century, causing alarm among whites and prompting in part the passage of the 1913 (Natives) Land Act which marked the beginnings of territorial segregation in South Africa (Simpson 1986: 84–96).

Many settlers had two farms each (about 3 500 hectares in extent, which could be held in freehold or quit rent), and they lived on the highveld in summer and spent the winter hunting on their lowveld or Waterberg farms. In the 1850s and 1860s hunting formed the basis of the Boer money economy, although agriculture in disease-free and more settled districts such as Potchefstroom and Rustenburg was growing. Recorded figures for agricultural exports during this period are indicative of farming activities in the region. For example, for the year January to December 1864 it has been calculated that exports from the Transvaal region yielded the following: ostrich feathers £25 000, ivory £30 000, leather £10 000, cattle £48 000, wool £30 000 – the feathers and wool coming from the southern region and the ivory from the Soutpansberg (Perry 1931: 35. See also J Carruthers 1995: 25–44; Cachet 1882; Huet 1869; Hofmeyr 1890 and Pelser 1950.).

In 1851 Natal trader John Sanderson visited the Magaliesberg and left an account of the month he spent there. He described a small community; Rustenburg was a church town (the site had been agreed on in 1848 at a meeting held on Kruger's farm, 'Hekpoort') with about fifteen to twenty houses, half of which were still being constructed.

Fig. 11.11 'The discovery of gold'. Painting by Thomas Baines (1874) commemorating Henry Hartley's discovery of gold in Zimbabwe while on a hunting trip

There was little wildlife in the immediate vicinity, but Rustenburg itself was a growing entrepôt for ivory and hides that came from further west and north – the Marico area, the Soutpansberg and Lake Ngami in present-day Botswana (Sanderson 1981: 247; see also Coetzee 1974). Despite the commercial activity, however, the annual revenue for Rustenburg in 1853 was £250 (McGill 1943: 34). The district was quite densely populated and becoming predominantly agricultural. Fruit and tobacco were the most common crops, although the fruit was subject to rust and blight and most of the peach crop was converted into a brandy known as 'Cape Smoke' – apparently a thriving industry. Sanderson skirted the Magaliesberg on its southern side, noting how the trees decreased in number as he neared the Witwatersberg (the ridge upon which the village of Magaliesberg was later established) and the road that led to Potchefstroom via Hol Fontein (Sanderson 1981: 233–255). A few years later, John Churchill, a trader who had been active in the Transvaal since the 1840s, also referred to the Magaliesberg as being thickly populated with farmhouses, and was admiring of the fruit that was grown and of attempts to cultivate coffee (Transvaal Archives TAB A17; Child 1979: 74–76). These traders met some of the African inhabitants of the area, Sanderson being particularly impressed by Magato and his people near Rustenburg, whose huts were beautifully built, and by the fact that 'the cleanliness pervading these native kraals is such as ought to shame the Dutch Boers' (Sanderson 1981: 249).

During the period of governmental instability before the Sand River Convention was signed, the Voortrekkers were suspicious to a paranoic degree of all visitors and parti-cularly of anyone who was English-speaking, because they were so afraid of losing their independence, as had happened to them in Natal. In April 1849 the trader James Chapman was detained at Mooi River and told to his 'infinite surprise' by the local landdrost, Hermanus Lombard, that he was not welcome to live north of the Vaal River. Chapman solved the problem by delivering two cases of cognac to the official the next day and was then allowed to lease a house and to obtain an extended hunting licence (Chapman 1971: 18–24). Henry Hartley, an English trader from the Eastern Cape, had settled in the early 1840s on the farm 'Thorndale' in the 'Cradle of Humankind'. He was a hunter by profession and in 1846, while en route to the Soutpansberg, he was arrested at Ohrigstad despite having a letter of introduction from his local commandant, Gerrit Kruger, that allowed him to trade as far north as the Olifants River (Transvaal Archives TAB SS0 R114/46). Also in 1846 and again in 1850, trader Joseph McCabe transgressed the Boer prohibition on any foreigner hunting in what was regarded as Boer territory. He was heavily fined and his goods were confiscated. McCabe apparently knew the way to Lake Ngami and it was feared that his map of that area would attract more English hunters into the Transvaal. Thomas Baines, too, was detained at Potchefstroom in 1850 and forbidden to make maps or sketches (Kennedy 1964 2: 58, 67–68). In 1852 Henry Gassiott, a hunter-visitor to South Africa, was also regarded as too inquisitive a foreigner, perhaps even a British spy. Gassiott records the Boers' suspicion of maps and map-makers in particular, and explains that the officials at Warmbaths (Bela Bela) refused to give him permission to travel further north. He ignored this injunction, visiting Ohrigstad but avoiding Soutpansberg, 'the furthest settlement of the Dutch Boers, who would in all probability have detained [me]' (Gassiot 1852: 137–140).

The Boers seem to have been less afraid of visiting naturalists, perhaps because many of them were not English. Johan Wahlberg, collecting for the museum in Stockholm in the early 1840s, was a friend of the Boers, having surveyed and laid out the town of Pietermaritzburg, the Voortrekker capital in Natal. Wahlberg (who was later trampled to death by an elephant at Lake Ngami) carried letters of introduction to Hermanus Potgieter in the Magaliesberg area from a relative in Natal, and was thus welcomed by the white community there. Wahlberg thought Potgieter's garden quite beautiful, and the district enhanced by the zebra and other game that he encountered on the road between Potchefstroom and the Magaliesberg that went via Wonder Cave and Hol Fontein. On the northern side of the Magaliesberg Wahlberg visited the camp of Karl Zeyher who, together with Joseph Burke, was collecting plants and animals for the Earl of Derby, Lord Stanley, president of both the Zoological and Linnaean Societies. Potgieter was not pleased about Wahlberg scouting around the Magaliesberg, but allowed him to go to the Apies River, the site of present-day Tshwane. On the way there Wahlberg bartered coffee for locally made shoes (*veldskoene*) while noting how many Africans brought tusks to Potgieter as items of barter. Wahlberg himself did not lack for African guides and helpers and as the hunting was good – plains antelope, sable and much else, a veritable 'earthly paradise' as he called it – there was no shortage of food for the party (Craig et al. 1994: 73–74, 154).

Delegorgue also experienced life in the Magaliesberg at this time. He found that at first the Boers simply would not believe that he had no political motivation and that he was visiting the Transvaal merely to collect and skin birds. Because no African guides would travel with him from Natal to the highveld – they knew how cold the winters were – Delegorgue relied on local Africans who told him that they did not want the Boers to know that they had formerly been part of Mzilikazi's community because they would then be dreadfully ill-treated out of revenge (Delegorgue 1997: 155, 162, 168).

New frontiers, new political powers
In the 1830s and 1840s the area that has been declared the Cradle of Humankind World Heritage Site was an open frontier. Race relations were fluid and no one group was able

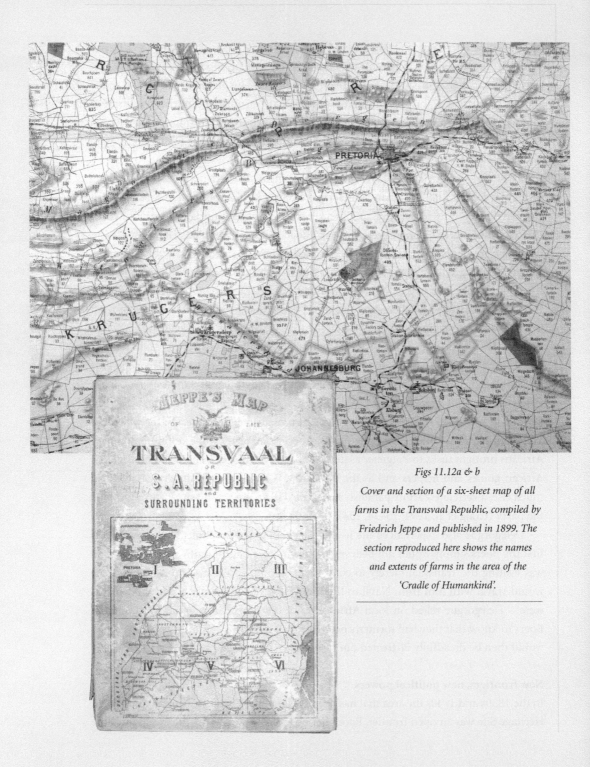

Figs 11.12a & b

Cover and section of a six-sheet map of all farms in the Transvaal Republic, compiled by Friedrich Jeppe and published in 1899. The section reproduced here shows the names and extents of farms in the area of the 'Cradle of Humankind'.

completely to dominate another. There was an abundance of wildlife that many travellers, traders and hunters considered unparalleled in world terms. Knowledge about the geography, agricultural potential and mineral wealth of the region was extremely scanty.

But within little more than five decades the frontier had closed. The social landscape had altered completely and with it the balance of political and economic power. White supremacy was secure, with much of the political leadership of the Republican Transvaal coming from the Boer community who lived in the Cradle area and its surrounding districts. A polity independent of Britain had prospered with agricultural development, and the later discovery of rich sources of valuable minerals had increased the wealth of the state and its white citizens. The whole of the Magaliesberg area had been divided into farms and parcelled out to settlers, most of them Boers. At first white farmers tended their crops during the summer and in winter migrated to the hunting grounds of northern South Africa and beyond. Before long, however, even the wildlife of the 'Far Interior' was decimated and not worth the trek. A denser white population soon began to show divisions within itself and a poor white class of *bywoners* or tenant farmers appeared in the Cradle area. Economic dominance was predicated on African labour, which was obtained by subjugating the Tswana communities, destroying the fabric of their society and forcibly wresting their land from them. In time the local people were transformed into units of labour and were often extremely badly treated, but their resistance was continuous and murders of whites were not rare occurrences (Allen 1979: 106, 115–116, 146, 161–162). By the end of the nineteenth century the social, intellectual, political and economic landscape of the Cradle area had been drastically transformed.

Early mining

PART 5

THE RACIAL PARADOX: STERKFONTEIN, SMUTS AND SEGREGATION

Philip Bonner

The gold discoveries on the Witwatersrand changed the face of South Africa. By 1896, mines on this reef were producing one-third of the entire world's gold supply and by 1914, two-thirds. Gold mining's voracious demand for cheap African labour partly authored the system of territorial segregation (which later developed into apartheid) with which South Africa later became synonymous. The demand for agricultural goods on the Witwatersrand also stimulated an economic revolution in agriculture, which prompted commercialising white farmers to expel landless Boer *bywoners* (tenants) from their farms, thus making visible for the first time the 'poor-white' problem which would dominate white politics in South Africa for the first half of the twentieth century.

The quest for gold in South Africa in the mid- to late nineteenth century is recounted in Chapter 12 by Philip Bonner, and emphasises the close connections of the Cradle area and of the palaeontological discoveries for which it later became famous, with the mining of gold. The first serious efforts at mining gold in the Witwatersrand area took place in the Cradle in the early 1880s; the remains of several of these mines are still visible today. The Cradle's palaeontological riches only revealed themselves as a result of the mining of lime, one of the prerequisites for the refining of gold.

The riches yielded by gold also greatly increased the size and significance of the geological and wider scientific community on the Witwatersrand. One result of this development was interest registered by successive South African governments and statesmen in exposing and promoting South African scientific endeavours, which they saw not only as economically advantageous but also as a marker of white South Africa's progress and 'civilisation'. The establishment of the University of the Witwatersrand in 1922 was one landmark in this process. The establishment two years later of a Chair in Anatomy, whose first incumbent was Edward Philip Stibbe, followed shortly thereafter by Raymond Dart, created the context in which some of the fossil remains exposed in the process of lime quarrying would be recognised as those of hominids, and humankind's earliest ancestors could be located in the sub-continent rather than in Europe or Asia.

Phillip Tobias, in Chapter 13, traces the multiple steps between the discovery and excavation of the Kromdraai and Sterkfontein lime deposits and caves, and the uncovering and comprehending of their rich fossil accumulations, culminating in Broom's

recovery of an adult australopithecine in 1936. In his meticulous and evocative reconstruction he succeeds in conveying not only the wonder of Sterkfontein but also the
fascinating, colourful and sometimes idiosyncratic characters of some of the scientists
who made such spectacular palaeontological advances.

The mineral discoveries of the late nineteenth century in South Africa also attracted
the interest and engagement of the European powers, most notably Britain. The discovery of diamonds at Kimberley around 1870 led Britain to try to create a Confederation
of South African states under British sovereignty, and to undertake the first British
annexation of the Transvaal. As part of this exercise, imperial British forces crushed the
most important surviving independent African chiefdoms such as the Zulu and the Pedi
in 1879–80, thus paving the way for the first time for uncontested white supremacy. The
venture of the Confederation had failed by 1881, but ongoing British interest in the sub-
continent led to the annexation of Bechuanaland (now Botswana) and southern
Rhodesia (now Zimbabwe). This effectively sealed off the expanding South African
frontier. Landless Boer hunters and herdsmen no longer had open to them new land to
appropriate, on which to make their fortunes. Within South Africa's borders the remnants of its teeming herds of wild animals were now shot out. This again accelerated the
growth of a poor-white class in South Africa. Ironically, the decisive attainment of white
supremacy almost instantly created its own Achilles heel, a class of whites who inter-
mingled with and might acculturate to blacks.

Gold, as noted already, also steadily shut down the internal frontier. The great
mining houses of the Rand bought up huge tracts of land for mineral and speculative
purposes and turned white *bywoners* off the land to make room for more lucrative rent-
paying blacks. Simultaneously, commercialising Boer landowners did the same. As
Beinart and Coates write, both in South Africa and in the USA in more or less the same
period, 'the frontier closed with the transition to intensive agriculture and the barbed
wire fence' (Beinart & Coates 1995: 12).

A mortal blow was delivered to the prospects of white *bywoners* by the South African
War of 1899–1902. The war was ultimately the product of the changes and rivalries set in
motion by the massive revenues yielded by gold, and turned out to be perhaps the most
formative event in South African history. During its guerrilla phase, which opened in late
1900, the British adopted scorched-earth tactics, destroying farmhouses, crops and stock
throughout the Transvaal and the Orange Free State. Tens of thousands of rural whites
were displaced, concentration camps were constructed in which Boer civilians were
placed at enormous cost of life and, to the outrage of Boer opinion, Africans were extensively used in defensive blockhouse lines and for scouting. Once the war was over many
bywoners were never able to reclaim a toehold on the land, and a festering bitterness
against the British and the Africans began to develop in sections of the Afrikaner

or Boer community. To this the real beginnings of Afrikaner nationalism can be traced.

One of the principal theatres of the guerrilla phase of the war, as Vincent Carruthers shows in Chapter 14, was the Cradle/Magaliesberg area. Perhaps the most celebrated Boer general, de la Rey directed operations in the Magaliesberg area and parts further west. As Carruthers demonstrates, many sites of epic battles can still be identified today, while several of the remaining blockhouses used by the British to anchor defensive barbed-wire lines are to be found there – some of the few that still survive in South Africa. Closer to Krugersdorp, the sites of two large concentration camps for Boers and blacks can be found.

The South African War of 1899–1902 was seen by most sections of the white population in South Africa as a serious blow to white supremacy. In nearby Derdepoort to the west, a Boer force had been massacred by the BaKgatla in the initial stages of the war, at least partly in revenge for the earlier expropriation of their land. During the war Africans raided Boer cattle extensively, and reassumed *de facto* occupation of the lands they had lost. Many Boer farmers feared returning to their land for several years after the war. Africans who entered British service during the war also earned substantially better wages than those paid on the mines and the farms. With this income, after the conclusion of war, they resumed the buying of land, as the BaPô had already been doing for several decades just west of Wolhuterskop on the doorstep of the Cradle. After the war, and particularly after the Act of Union in 1910 which restored independence to white South Africa, South Africa's white politicians, among whom Smuts and Hertzog were prominent, began attempting to reverse the process whereby poor whites were sliding down the economic and social ladder, and slightly more prosperous blacks were climbing up. The overarching solution which they proposed was territorial and residential segregation. The 1913 Land Act proclaimed that 82 per cent of the Union land was reserved for whites. Piecemeal black encroachment was stopped. The Cradle/Magaliesberg area was guaranteed to remain white.

Nevertheless, the poor-white problem did not disappear. Rather, it grew as drought, white population increase and agricultural commercialisation remorselessly squeezed the poorer sections of the white population. Alarm mounted and countless initiatives were undertaken to 'save' and 'rehabilitate' this section of the population. The *Report of the Select Committee on European Employment and Labour Conditions* of 1913 expressed a view, typical of those at the time, which was articulated repeatedly thereafter:

> The importance of the [poor white] question in South Africa arises from the fact that the European minority, occupying as it does, in relation to the non-European majority, the position of a dominant race, cannot allow a considerable number of its members to fall below the level of the non-European worker. If they do…sooner or later, notwithstanding all our material and intellectual advantages, our race is bound to perish in South Africa (cited in Berger 1982: 58).

To avert the threatened challenge to white supremacy, the Smuts and successive governments initiated a range of relief works (roads, dam building, etc.) and 'civilised labour policies' which reserved unskilled jobs for whites (for example on the railways) but at higher rates of pay than those enjoyed by unskilled blacks. Short-term relief works were deemed less satisfactory, as they offered few long-term possibilities for rehabilitating and disciplining whole families, particularly placing children in schools: long-term relief would equip whites to enter the labour market in a privileged position compared to blacks.

One poor-white relief project which promised to combine both aims was the Hartbeespoort Dam. The Hartbeespoort Dam lies at the door of the Cradle. Today it offers a leisure world of water sports to a middle-class and largely white, suburban population working and residing in Pretoria and Johannesburg, its poor-white origins forgotten and relegated to a distant past. As Chapter 15 by Tim Clynick shows, however, the Hartbeespoort Dam project was conceived of during the First World War as the flagship poor-white initiative in South Africa. After construction started in September 1918, the number of whites employed on the project grew rapidly until they numbered 3 400 in 1922. The dam project was perfectly suited to the government's rehabilitatory objectives, since its long-term nature allowed families to join the workers in family accommodation, and proper schooling could be provided for children. At its height a white population of 10 000 was supported by the dam. Ironically, part of the housing provided for white workers was constructed on Wolhuterskop (it still stands there awkwardly today), the ancestral home of the BaPô (see Chapters 10 and 11 by Simon Hall and Jane Carruthers, in Part 4). After the dam was completed, other rehabilitative aims of the government were realised, as Clynick shows, in the Losperfontein irrigated land settlement for whites which offered a framework of disciplined, 'civilised' living for erstwhile 'poor whites'.

The Cradle and the Hartbeespoort Dam stand confronting one another. Each in its own way is a monument to JC Smuts. Smuts was as alarmed as any other white politician at the threat posed to white supremacy by poor whites, and so sponsored the Hartbeespoort Dam project. At the same time, he actively promoted the palaeontological work of Raymond Dart at the University of the Witwatersrand Medical School, and endorsed the scientific validity of Dart's findings that the Taung skull represented evidence of the evolution of humankind – or as he would have put it, the human race – in South Africa. In 1936, in another paradoxical synchronicity, Smuts celebrated Robert Broom's hominid find at Sterkfontein, which confirmed Dart's earlier claims for the Taung skull, while supporting the Hertzog Bills which finally elaborated the policy of segregation. At face value, these two attitudes and assumptions sit incongruously and anomalously together. On the one hand, Smuts was a prime architect of the white supremacist system of segregation which asserted the inferiority of blacks. On the other hand, he supported the view that the human race had evolved in South Africa and not in 'civilised' Europe and Asia, as most other scientists of the day proposed and preferred. These attitudes can be partly reconciled, as Dubow (1995) suggests, by interrogating Smuts's understanding of civilisation. For him a prime attribute and diagnostic feature of Western civilisation was Western science. Smuts promoted and wished for a South African science as evidence of white South Africa's higher civilised status, and as a core and unifying component of a white identity and white race. In addition, as Dubow shows in Chapter 1, Smuts subscribed to the view that a linked evolutionary past did not necessarily imply a belief in a common humanity. Rather, humankind was composed of different kinds of humans. Even so, the connections of civilisation with Later Stone Age South Africa were uncomfortably close. He ultimately managed a partial resolution of his intellectual dilemma in a neat racial sleight of hand. First he posed the question:

> As they were racially and physically not very different 15 000 years ago, what has caused the immense difference between the European and the Bushmen of today? We see in the one the leading race of the world, while the other, though still living, has become a mere human fossil verging to extinction (Schlanger 2002: 9).

The answer that he then offered was crude but straightforward. While part of the original population undertook a tempering and civilising trek north (this last phrase was Schlanger's), the Bushmen remained in place, 'shrivelling' in the South African sun. Several thousand years later the early migrants to South Africa completed the round trip. As Smuts observed:

> …the Europeans (descendants of those earliest migrant cousins) who now *return*…find a very different situation from that which their *African ancestors* left some 15 000 years ago (Schlanger 2002: 9).

There remained nevertheless a flaw and inconsistency in Smuts's thinking, as a man of his intellectual range and accomplishment must have at least momentarily and privately recognised. If whites were collectively endowed with such powers of civilisation, why did they need so much artificial protection to lift them out of the abyss of poor-whitism and maintain the monolith of white supremacy? In another context Smuts would leave the resolutions of such problems to 'the broader shoulders of future generations'. It seems fair to claim, therefore, that he, more than any other figure, embodies and personifies the paradox that is Sterkfontein – emblematic as it is of the complexities, the contradictions and the delusions of old South Africa. It is wise to recall once more President Mbeki's comment in his Maropeng speech, 'Africa is seldom what it seems.'

CHAPTER 12

THE LEGACY OF GOLD

Philip Bonner

Gold explains why South Africa was the site of the first early hominid fossil discoveries in Africa. Gold also explains why South African palaeontologists have been at the fore-front of palaeontological research ever since the discovery of the Taung skull by Raymond Dart in 1924.

Gold changed the face of South Africa. The massive reserves of gold uncovered on the Witwatersrand in the 1880s and 1890s led to the South African War of 1899–1902, and the construction of a modernised, centralised South African state. The economic dynamo of gold likewise transformed South Africa into the most developed economy on the continent of Africa. Finally, the discovery of gold gave South Africa's scientific com-munity a massive intellectual transfusion, vastly expanding the scale of geological research in the country, and prompting at one remove the birth of the science of palaeontology. Without gold the incredibly rich fossil heritage of Sterkfontein, Swartkrans and Kromdraai would have taken far longer to be discovered, and probably longer still to be apprehended. This chapter traces the story of the discovery of gold in the Witwatersrand area, and more specifically in the area now known as the 'Cradle of Humankind', and the subsequent union of science and gold in the late nineteenth and early twentieth centuries.

Scientific exploration in nineteenth-century South Africa

A number of scientific explorers ventured into the interior of South Africa in the early to mid-nineteenth century, probably because the substantial port town of Cape Town offered an ideal springboard for such expeditions. Andrew Smith, whose exploratory activities in the region have been discussed in Chapter 11, was one of the first travel-lers to record the ruins of the immense Tswana cities on the doorstep of the Cradle area, in 1832; he did so while engaged in a geological and fossil-hunting expedition, and his route seems to have taken him within a short distance of the Sterkfontein deposits. At more or less the same time hunters and would-be colonists were fanning out through the interior of southern Africa, as described in Chapter 11. One such adventurous individual, by the name of Karel Kruger, claimed to have discovered gold ore somewhere along the Witwatersrand in 1834, and took samples back to the Cape. SP Erasmus and three of his sons accompanied Kruger on a second visit in this direc-tion in 1836. After crossing the Vaal River and proceeding in the direction of

Sterkfontein, they were attacked by a party of Matebele warriors in one of the first clashes between black and white parties, presaging the conflicts which would intensify as groups of Voortrekkers penetrated the land beyond the Cape Colony in what became known as the Great Trek. Kruger and most of the party were killed. Only Erasmus and one of his sons survived.

Word of potential riches in gold in the interior began to circulate widely over the following two decades, causing this region slowly to become part of a globalising culture of gold prospecting and mining. From the late 1840s, gold rushes were bursting out in various parts of the new and the colonial world. Gold prospectors and gold geologists circled the globe in the feverish quest for the valuable ore, moving from California to Australia to southern Africa and many other parts. One such figure drawn to the Witwatersrand by the allure of gold was Pieter Jacob Marais. Marais was born in Cape Town in 1826. In 1849 he set out to join the California gold rush (taking out US citizenship in the process). In 1852, following news of the discovery of gold at the Bendigo deposits in Victoria, Australia, he set sail for those distant shores. A year later he returned to South Africa, where he continued his search for the precious metal in the interior.

Working his way along the Crocodile River, Marais found a small quantity of alluvial gold on a nearby farm. He then followed the tributaries of the Crocodile River into what is now Sandton (part of Johannesburg), discovering a little gold in the Braamfontein Spruit on the farm 'Zandfontein'. Marais reported his finds to the Boer authorities and they insisted that the matter be kept absolutely secret (Coates 1987: 31–33). He was rewarded by the legislature of the Potchefstroom Republic with a concession to prospect for gold. News of Marais's discoveries leaked out, and was published in the Free State newspaper *The Friend* in February 1854.

Other, more formally qualified geologists followed in his footsteps, most notably the Germans Carl Mauch in 1866, Emil Holub in 1871 and E Cohen in 1873, each of whom added significantly to the corpus of scientific knowledge being gathered together in South Africa, and the public esteem in which it was held. In the 1860s and 1870s, the interest of prospectors and geologists drawn from all parts of the world was attracted to the headwaters of the Limpopo River and to the gold diggings in Pilgrim's Rest and Barberton (in the province of Mpumalanga, then known as the Eastern Transvaal). Finds nevertheless continued to be made in the wider Sterkfontein area, which continued to be the main centre of interest of prospectors and geologists alike in the highveld area of the Transvaal (now Gauteng province) until the discovery of the Main Reef on the Witwatersrand in 1886. Substantial deposits of gold were found right in the Cradle area in 1874 on the farm 'Zuikerboschfontein' (later called 'Blaauwbank') by Australian digger Henry Lewis, and JH Jennings and J Jennings, the owners of the farm (Wallis

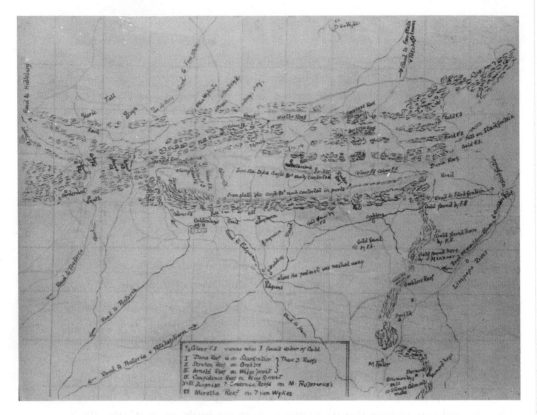

Fig. 12.1 Struben's gold-prospecting map of the Witwatersrand

1946: 18 *et passim*). This find was reported to the Transvaal government in terms of the Gold Law of 1871, which made the reporting of all such finds compulsory. The government eagerly anticipated its share of the profits; by this time, the Transvaal government under President Thomas Burgers was deeply in debt. Products of the hunt were dwindling, agriculture was slow to take off and the ongoing conflict with African groups in the region took its financial toll. Gold mining was considered to be the only way of avoiding looming bankruptcy. 'Blaauwbank' was declared a public diggings in January 1875 and the Nil Desperandum company started operations on the farm. The returns were, however, disappointing.

Fig. 12.2 Prospectus of the first Witwatersrand gold-mining company, Nil Desperandum

Gold and lime

In 1881 the first major payable reef in this area was discovered and worked by SI Minnaar on the farm 'Kromdraai'. In 1884 shafts were sunk by the well-known gold prospector JG Bantjies on the farms 'Kromdraai' and 'Uitkomst', and in 1884–85 the neighbouring farm 'Tweefontein' yielded 750 pounds of gold. The remains of some of these mines, such as Black Reef, can still be seen in the Sterkfontein area.

Meanwhile, and ultimately more importantly, HW and FFT Struben had undertaken the earliest serious prospecting on the farm 'Sterkfontein' in January 1884. Gold had first been reported on 'Sterkfontein' in

Figs 12.3a & b Letter from Minnaar and Lisemore requesting permission to prospect, and translation of letter

R 420/82.
Pretoria,
13th January, 1882.

MR. EDWARD BOK,
State Secretary.
DEAR SIR,
We request your Honour to issue to us a prospecting licence for the farms named Kromdraay and Honig Klip, Driefontyn and Rietfontyn, each adjacent to the other, all in the Pretoria district. Kromdraay, H. J. Grobler. Honig Klip, H. F. Grobler. Driefontyn, H. Mulder, Snr. Reitfontyn, D. P. Pretorius. We desire to have all of the said farms for an unspecified period, not less than six months. Should we, the undersigned, succeed in our expectations we trust that the Government or the State Secretary will grant a *Concession* on proof of the discovery of mineral lodes such as gold, silver and lead or any substances.
We trust that your Honour will not reject our petition.
We remain,
Your Honour's
Obedient servants,
S. MINNAAR.
T. LISEMORE.

209

mid-1878, but not in payable quantities. On his second day of prospecting in January 1884, Fred Struben found gold-bearing reef. The following month, the Strubens purchased the farm 'Sterkfontein' (its owner, OA Jacoby, having become insolvent in 1883) and formed the Sterkfontein Junction Gold Mining Company. On this rested Struben's much vaunted claim to have been the discoverer of the goldfields in the Witwatersrand District. Thenceforth his attention turned in that direction.

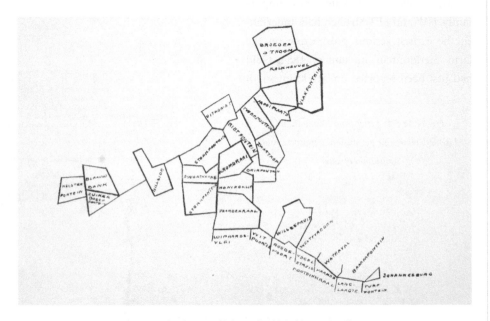

Fig. 12.4 Sketch map of 'Blaauwbank' and 'Kromdraai' area

Up until 1883, prospectors had overlooked what was soon to be revealed as the main source of gold, the banket reef or conglomerate ore. 'Banket' was an Afrikaans word for hard, round, baked sugarball sweets. Banket reef comprised round and almond-shaped alluvial pebbles, cemented together in silica and sand and thickly sprinkled with pyrites. Traces of banket gold were found by various prospectors on the Witwatersrand in about 1883, but in reefs that were too thin to produce payable gold. The Main Reef of gold-bearing banket or conglomerate was not discovered until March–April 1886 on the farm 'Langlaagte', which abutted the farm 'Braamfontein'. Different accounts contest the individual who discovered the Main Reef. Pride of place usually goes to the Australian prospector George Harrison, or otherwise to the Englishman George Walker. Both were working on the farm when the find was made. Others attribute it to Louis Oosthuisen, whose father owned that portion of 'Langlaagte', and Koos Ackerman.

THE STAR, JOHANNESBURG, TRANSVAAL, FRIDAY, FEBRUARY 6, 1925.

ANTIQUITY OF MAN.

KNOWLEDGE REVISED.

THE SOUTH AFRICAN REVELATIONS.

MAN AND APE FROM SAME STOCK.

FROM OUR OWN CORRESPONDENT.

LONDON, Thursday. — Sir Arthur Keith's famous book, "The Antiquity of Man," which was issued to-day in a new and enlarged edition, has particular interest in view of Professor Dart's discovery of the Taungs skull.

Sir Arthur Keith declares that "an unexpected chapter in the history of mankind" is being revealed in South Africa. He correlates the deposits and implements previously found in South Africa with those found near Ipswich, England, and states we must infer that "both South Africans and East Anglicans were subject to various cultural waves which slowly crept over the world from time to time."

Human existence, in Sir Arthur Keith's view, may reach into the dim geological ages. "There is not a single fact known to me," he writes, "which makes the existence of the human form in the Miocene period an impossibility," in which case the age of man may have to be measured by millions of years.

Well-informed comment on Professor Dart's discovery points out that it does not imply the direct descent of man from the ape. The "Daily Telegraph" says: "The origin of the human species must be sought far back in the world's development. Man was not evolved from the gorilla or chimpanzee, but they and he all sprang from the same stock at different periods to grow in different ways and adapt themselves to different purposes by different means."

PROFESSOR DART.

GEN. SMUTS'S THANKS FOR SERVICE TO S. AFRICA.

General Smuts has addressed the following letter to Professor Raymond Dart, of the Witwatersrand University.

Dear Professor Dart : I wish both personally and as president of the South African Association for the Advancement of Science to send you my warm congratulations on your important discovery of the Taungs fossil. Your great keenness and zealous interest in anthropology have led to what may well prove an epoch-making discovery, not only of far-reaching importance from an anthropological point of view but also well calculated to concentrate attention on South Africa as the great field for scientific discovery which it undoubtedly is. The recognition of the unique importance of our Rhodesian Broken Hill skull in human evolution has now immediately been followed by your discovery, which seems to open up a still further vista into our human past. I congratulate you on this great reward of your labours, which reflects lustre on all South African science, and I wish to express the hope that many further triumphs await you and those who have so willingly co-operated with you on the road on which you have begun so well.—Yours very sincerely,
J. C. SMUTS.

TAUNGS LIMESTONE.

HOW THE DEPOSIT WAS DISCOVERED.

DURBAN, Friday.—Your correspondent has had an interesting chat with Mr. M. G. Nolan, of Durban, who was the actual discoverer of the huge bed of limestone wherein the famous Taungs skull was found.

Mr. Nolan, who has made a special study of limestone deposits, is ordinarily engaged in business connected with music and musical interests, but as that business takes him up and down the country a good deal he is able on occasion to indulge his fondness for prospecting.

"My brother," said Mr. Nolan, "is the owner of the Nolan lime works in Sterkfontein, near Krugersdorp. He commissioned me in my wanderings about the country to keep my eyes open for any considerable deposits of limestone. Eventually I landed at Taungs, and while looking towards the east along the Kaap Plateau I noticed a white formation at about seven miles distant. Inquiries among the people living round about revealed the fact that this was dolomite. Knowing, of course, that dolomite is the mother of all limestone, I came to the conclusion that this white mass in the distance would most probably turn out to be the sort of thing I was looking for.

"Following a ride on my bicycle, which I left in the charge of friendly natives at a neighbouring kraal about a mile away, I made my way through dense bush to my goal, and found what I verily believe is the greatest deposit of limestone in the world. I lost no time in procuring a claim licence at Vryburg entitling me to take over 1,000 acres. I then went back to Taungs and located the scene of my find, a place which the natives call Thaba Sige ("Black Mountain"). From rough measurements I concluded that the area under limestone was approximately the area allowed me under my licence. It was in this great bed of limestone, at a depth of fifty feet below the surface, that the Taungs skull was found. While that unique relic is of enormous interest to the world of science the great limestone deposit is of more material value."—Special.

Fig. 12.5 News cuttings from The Star, *6 February 1925, in which Smuts thanks Dart at the time of the Taung skull find*

Gold provided the material resources, the physical infrastructure and the social scaffolding which supported the growth of a serious scientific community in South Africa. A generation of leading South African scientists took up geology as a profession on the Witwatersrand in the wake of the discovery of gold. By 1890 they were sufficiently numerous to encourage the formation of the Geological Society of South Africa, which was responsible for recognising the first fossil-bearing breccia at Sterkfontein. In 1910 the Transvaal School of Mines in Johannesburg was elevated to the status of a university college and with the powerful backing of the Witwatersrand's mining interests, a fully-fledged university, the University of the Witwatersrand, was established in 1922, thereby realising Milner's and Chamberlain's dream of founding 'a Scientific university (in Johannesburg) specialising according to the needs of the great industries of the community' (Murray 1982: 1, 11–12). An anatomy chair was already in existence at the former university college, and one of the founding facilities of the new university was the Medical School. In 1924 Raymond Dart took up a position as head of the university's Department of Anatomy, thereby creating the link between the world of science and the world of mining and geology which would shortly afterwards present the wider sci-

entific community with the world-shattering discovery that Africa, rather than Java, Peking or Piltdown, was the home of humankind's hominid forebears.

Gold was also instrumental, in a rather different way, in creating the physical context in which palaeontology and the study of fossils would germinate and then flourish. The Macarthur Forrest cyanide reduction process, which was the lifeline for the deep-level Witwatersrand gold mines, required inputs of massive quantities of lime. A distinctive feature of the geology of the inland highveld/bushveld areas of the region known as the Transvaal was that it was straddled by innumerable limestone caverns. From the early 1890s on these were dynamited and excavated on a massive scale. In the breccia uncovered, removed or left behind lay immured the fossils of innumerable ancient species of animals, insects, plants and, most famously, hominids. It was thus the conjunction of fossil-bearing breccia exposed by lime workings in ancient caverns, and a scientific community geared initially to serving the interest of the gold mines, that installed South Africa as one of the premier players in the world of palaeontological research.

Both the Kromdraai and Sterkfontein caves were the sites of early lime quarrying, the former by Hans Paul Thomasset, an Englishman who travelled to South Africa in the early 1890s; the latter by Guglielmo Martinaglia, who had emigrated from Italy to South Africa in 1879, and who ended up on the Witwatersrand in 1886. In 1887 Martinaglia married a local Sterkfontein girl, Maria Magdalena du Plooy, and it was through her that he probably first became aware of the Sterkfontein cave deposits. He obtained a lease of the Sterkfontein lime deposits situated on the farm 'Zwartkrans', most probably after his sale of the Witpoortje Inn to James Richardson in 1893. Then, or some time shortly after that, Martinaglia entered into a business relationship with HG Nolan of Krugersdorp (Tobias 1983: 46–52). In the years that followed, business boomed. In its issue of 18 January 1895, *The Star* reported that the Krugersdorp Lime Company was 'developing rapidly'.

Fig. 12.6 Old mine workings at Kromdraai

Fig. 12.7 Dr Broom addressing the audience at the unveiling ceremony of a bust of him
at the Transvaal Museum on 31 October 1949. Seated on the left are Professor DE Malan
(Chairman of the Board) and General JC Smuts

'The Company's ground,' it went on,

is six miles from Krugersdorp, extends six miles by four and the company has the
sole right of mining thereon. The farm is very rich in limestone...[with] a plentiful supply of water. There is no need for making a shaft, as stone can be procured
by driving in the face of the [hill?] from which point it is run down into the kilns,
and thereafter being burnt it is run down into the sheds for the purpose of sorting and loading. Three kilns, one constantly burning, turning out 50 tons a week
of lime with a percentage of 95%...[are] used at Royal Cyanide Works... At
present half a dozen wagons are constantly moving between the works and the
Krugersdorp station and the demand exceeds the supply. To meet the demand
alterations are being made by the manager, MF Glass[!]. He intends constructing
a mill to manufacture cement [my additions].

By 1896 the Sterkfontein diggings had created a small quarry. Later that year or early in
1897, Martinaglia intensified his search for lime, and through blasting opened up a small
hole into the massive subterranean cavities. There, to quote Tobias (1983: 49), 'a veritable
fairyland of stalactite and stalagmite formations was laid bare.' A flurry of scientific
interest followed from the Geological Society of South Africa in the late 1890s and again
in the early 1900s (Malan 1959).

The first breccia yields

The earliest known fossils from the Sterkfontein area were identified in 1895 by the
geologist and founder member of the Geological Society of South Africa, David Draper.
In 1898 a group of French priests from the Marist Brothers School in Johannesburg
visited the recently exposed Sterkfontein deposits, in which they recorded a variety of
fossil bones.

As early as 1895, Draper mentioned that a part of a fossilised hominid skull had been
seen (but not preserved) in Kromdraai, and it is likely that fossilised bones were
removed from Sterkfontein by sightseers and others when it was being mined for lime.
The world had to wait until 1936 for the significance of these early finds to be recognised
when, inspired by Dart's identification of the Taung skull as the early hominid 'missing
link' twelve years earlier, Robert Broom discovered and broadcast to the world the
fossilised skull of 'Mrs Ples'.

CHAPTER 13

THE STORY OF STERKFONTEIN SINCE 1895

Phillip V Tobias

The story of the discovery of the Sterkfontein fossils begins in 1895. In that year Hans Paul Thomasset began quarrying for lime in the Sterkfontein area and especially in the Sterkfontein cave (Fig. 13.1).

About the same time, two other personalities with well-known names appeared on the scene. One was David Draper and the other Guglielmo Martinaglia; Draper's historical link was with the Kromdraai caves and Martinaglia's was with the Sterkfontein caves.

Early in the 1890s, Draper was elected a Fellow of the Geological Society of London, the first South African-born Fellow to be elected to that distinguished Society. Draper contributed a number of papers to the London Society. At the same time he realised the need for a local Society. Largely as a result of his efforts, coupled with those of Dr Hugh Exton, there was formed in 1895 the Geological Society of South Africa. At its inaugural meeting in February 1895, Draper was elected its first secretary and treasurer.

At the first ordinary meeting of the new Society, held on 8 April 1895, Draper made the following important contribution to the discussion:

> He had a short time ago visited the Kroomdraai caves [sic], or rather, the formations that once had been caves and he found them most interesting from a geological point of view. There was much to discover there. He, therefore, suggested that they make a beginning by making up a party to go and explore this interesting geological ground. They would make a thorough examination in four days, and the time would be well spent (Exton 1895: 10–11).

On 1 February 1895, Draper sent a box of bone-bearing breccia from Kromdraai to the British Museum of Natural History in South Kensington, London. The box of bones was accompanied by a letter dated 1 February 1895, part of which read:

> Mass of rock containing a number of fragments of bones. This is from a cave on the farm Kromdraai situated about 16 miles west of Johannesburg. There is a bed of stalagmite with masses of rock in which the bone is abundant. It is really a cave deposit containing bone breccia or fragments of bone. I send you the specimen for examination as I should like to know whether the rock is similar to that found in European caves. I am informed that portions of a human skull and the claw of a

lion were found in the cave but I have not seen the specimens nor have I yet inspected the cave though shortly hope to do so [original read by me in the British Museum of Natural History in September 1977 and cited by Oakley 1960].

From Draper's statement that he had not yet inspected the cave, we can place his visit some time after 1 February 1895 when he wrote to the British Museum, but before 8 April 1895, when he mentioned his visit to the Geological Society. What we do not know is who collected the box of bone-bearing breccia that Draper sent to London on 1 February 1895, nor from whom he obtained the information that a supposed human skull and lion claw had been seen in the cave.

No report on the fossils sent to the British Museum seems to have been received by Draper and the specimens remained there, apparently untouched, for sixty-three years. By a strange turn of the wheel, the material languishing among the stored collections of the Natural History Museum was 'rediscovered' and 're-excavated' there by none other than Dr LSB Leakey in June 1958. Dr KP Oakley, then head of the Sub-Department of Anthropology in the Museum, reported that they included remains of baboon, a small carnivore, a large horse, porcupine, rodents, a small lizard and birds.

Fig. 13.1 India ink drawing of the house occupied by Mr Hans Paul Thomasset on the farm 'Sterkfontein', winter 1896

Mar. 1, 1895   MACHINERY.   175

Natural Science.

Under this heading, a series of articles on
SOUTH AFRICAN COAL,
by Mr. David Draper, F.G.S., will be commenced in our
next issue.

In the December number of MACHINERY appeared a
paragraph, headed :—

ANCIENT TOADS.

After it was in print, the writer discovered the following in the weekly named *South Africa*, published on the 9th November, 1894 :—

"A strange discovery was made in the De Beer's mine recently.

"Two diamonds were found imbedded in a piece of well-preserved wood, found at a depth of 700 feet. On splitting the block, which was fully 10 inch. in diameter, a cavity was disclosed to view, in which reposed a living specimen of a tree frog. The local scientific men judge the age of the tree at 180 years, during 150 of which froggy had been entombed. The jumping creature did not long survive the shock of gazing upon a busy world after so long a seclusion, but soon expired. Figuratively speaking, it lives in a bottle of spirits in the Kimberley Museum."

This tale looks sufficiently circumstantial for belief, and there does not seem to be anything in the shape of a hoax about it, I thought.

Surely, after all, I have been wrong. Such things do occur, in defiance of the dictum of scientists.

I consequently wrote to the General Manager of De Beer's, sending him the cutting from *South Africa*, and received the following courteous reply:—

"I have to acknowledge the receipt of your letter; and, in reply, I beg to say that there is not, so far as we know, the slightest foundation for the story.

(Signed)   "GARDNER F. WILLIAMS
        "General Manager."

Evidently the Editor of *South Africa* has been taken in this time.

I have no doubt in my own mind that if the authentic (?) cases that are repeatedly being quoted about living frogs in rocks are investigated, they would be found, like the one of *South Africa*, inventions of imagination.

THE LIMESTONE CAVERNS OF THE TRANSVAAL.

Among the many processes of Nature whereby great results are achieved from apparently insignificant sources, is the forming of caverns by the gradual dissolution of the mineral constituents of the rocks by the action of the acids in rainwater.

Naturally, an exceedingly long period of time is required for the excavation of a cave such as that at Wonderfontein, which extends for many miles underground (its farthest limit having, up to date, not been reached), and which has a river flowing through it, probably for miles of its length.

The process is an exceedingly slow one, and is as follows:—Rainwater contains, amongst other constituents, a certain amount of carbonic acid. This attacks and dissolves the carbonate of lime in the rock, removing it in solution, and thus forming a cavity. Where the water percolating through the roof of this cavity is small in quantity, sufficient to form drops only, pendants of carbonate of lime, called stalactites, are formed. As each drop gathers on the roof and begins to evaporate and lose carbonic acid, the excess of carbonate of lime which it can no longer retain is deposited round its edges as a ring. Drop succeeding drop, the original ring forms into a long pendant tube, which, by subsequent deposits inside, becomes a solid stalk, and, on reaching the floor, may form a massive pillar.

Meanwhile, the larger streams of water which enter through the roof are removing, mechanically the more solid constituents of the rock, which become liberated by

original cavity gradually becomes a cave of great extent, opening out into large chambers where the rock was more easily acted upon by the water, and narrow passages where the rock was more durable or the quantity of water insignificant.

The continuation of the process for a great length of time leads to the weakening of the roof, until, eventually, it falls into the cave, and a great "sink hole" is seen on the surface.

In some instances, the quantity of stalactites has protected the roof from falling in, and they have formed pillars from floor to roof, and of sufficient strength to support the superincumbent strata.

These caverns are very numerous in the Transvaal, especially in the districts of Potchefstroom, Malmani, Pretoria, and Lydenburg.

At Krugersdorp, a number of them exist, and recently a quantity of bones, some of lions and antelopes, and a few apparently human remains, have been found on the floor, entombed with a hard rock formed out of carboniferous earth cemented together by lime into a compact mass. Unfortunately, what would have been a very interesting discovery has been completely destroyed by the hands of those who first found the animal remains, in their anxiety to show the specimens to their friends, and consequently a large quantity of these interesting relics of the past have been scattered about the country.

It is to be hoped that the Geological Society, when it gets fairly to work, will make an effort to explore the numerous caves in the country, and pay especial attention to the remains of those animals which sought shelter in them, or which were dragged in to sustain the lives of the carnivora which probably inhabited them.

The Huntington Centrifugal Roller Quartz Mill.

The Huntington Centrifugal Roller Quartz Mill is now in its thirteenth year of use as a competitor against all kinds of mills for crushing ore, and during that time has largely increased in favour with mining men in all mining parts of the world, and is regarded as the most successful competitor against stamps which has yet appeared. Its construction is very simple, and the makers have carefully avoided the fatal defects of previous pulverizers. It represents a combination of fast running roll, with the mortar and screening advantages of a stamp mill. The ore and water being fed into the mill, rotating rollers and scrapers throw the ore against a die-ring, where it is crushed to any desired fineness by the centrifugal force of the rollers as they roll over it. The water and pulverized ore are thrown against and through the screens when fine enough. The discharge is so perfect that it makes little or no slimes, and leaves the pulp in good condition for concentration. The rollers are suspended, leaving a space of one inch between them and the bottom of the mill, thus allowing them to pass freely over the quicksilver and amalgam without grinding it or throwing it from the mill, while it agitates it sufficiently to make amalgamation perfect. One of the chief claims for merit in this mill is the prevention of all flouring of gold and quick-silver, and the consequent loss of gold that attends stamp-milling.

This mill is excellently adapted for this country for many reasons, chief amongst which are freight, which, as is well-known to be very heavy items here, the Huntingdon mill can be freighted and erected at one-eighth the cost of stamps, and requires no skilled labour for erection, which advantages in themselves give these mills a great recommendation to those who are starting to work claims with a small capital in hand. Again, very much less power is required to run a mill of this kind, which, in itself, is another great saving. A great many of these mills are in use in the States, where they have been found most useful in starting preliminary work on a mine. Messrs. Fraser and Chalmers, of Chicago, are the makers of this useful and well adapted little mill, of which full

Fig. 13.2 Article in Machinery *on the discovery by G Martinaglia of the cave system at Sterkfontein in 1896*

Draper's discoveries were made public in the 1 March 1895 issue of the journal *Machinery,* which contains an unsigned article headed 'The Limestone Caverns of the Transvaal' (p. 175). The article discusses the mode of formation of the caves and of the stalactites and stalagmites, and adds, 'These caverns are very numerous in the Transvaal, especially in the districts of Potchefstroom, Malmani, Pretoria and Lydenburg.' Then follows this most important passage:

> At Krugersdorp, a number of them exist, and recently a quantity of bones, some of lions and antelopes, and a few apparently human remains, have been found on the floor, entombed with[in] a hard rock formed out of carboniferous earth cemented together by lime into a compact mass. Unfortunately, what would have been a very interesting discovery has been completely destroyed by the hands of those who first found the animal remains, in their anxiety to show the specimens to their friends, and consequently a large quantity of these interest-ing relics of the past have been scattered about the country.

The date was 1 March 1895. It was exactly one month since Draper had written to the British Museum with the box of bones from Kromdraai. It is highly likely that the arti-cle in *Machinery* was referring to the Kromdraai bones and that Draper was the author (Fig. 13.2).

At the second meeting of the South African Geological Society on 13 May 1895 Draper delivered a wide-ranging paper on the conglomerate beds of the Witwatersrand (Draper 1895), and the dolomite rocks out of which the Bloubank River Valley was formed. In this address Draper referred to the numerous caverns, underground water-courses, and sinkholes which occur in the dolomites, formed by the solvent action of water. He added:

> That many of these caverns have lately been inhabited by wild animals has been demonstrated by an examination of the earthy matter which formed the floor of these caves, and which has yielded a large quantity of animal remains...they only show that these caves are of comparatively ancient date, and that they formed the place of refuge for animals existing at that time (Draper 1895: 20).

This general statement does not relate, in Draper's paper, to any specific cave; but it would seem to be one of the very earliest references to South African cave breccias con-taining bones.

The role of Guglielmo Martinaglia

The famous underground caves of Sterkfontein, with their fantastic wealth of stalactites and stalagmites, dolomite and chert formations, were blasted open by Guglielmo Martinaglia (Figs 13.3 and 13.4) sometime between the end of the Jameson Raid (on 2 January 1896) and the monthly meeting of the Geological Society on 12 July 1897, when David Draper announced the 'recent discovery' of the extensive system of underground caves (Draper 1897; Malan 1959).

Sometime in 1893 Guglielmo Martinaglia had obtained a lease on the lime deposits on the farm 'Zwartkrans'. Martinaglia was working in partnership with Nick Nolan of Krugersdorp. His son, Dr Giovanni Martinaglia, later wrote:

Figs 13.3a & b Guglielmo Martinaglia and his first wife
Maria Magdalena du Plooy, with a covering note by Phillip Tobias

There is little doubt that the fullest co-operation existed between my father and the late Mr HG Nolan in connection with the prospecting which was taking place, but on the day when the sensational Stalactite Caves were blasted open on the wall of the white quarry, my father was in sole charge of the operations. I remember this well, as I paid daily visits to the scene at which he was working (Martinaglia 1947; Tobias 1983).

As Dr Martinaglia recorded elsewhere:

My father was prospecting for lime and one day when I took him some refreshments to the quarry where he was working, he pointed enthusiastically to a dark hole in the side of the quarry and told me in Afrikaans that he had just shot open a 'wondergat'. This made a great impression on my boyish mind and I still remember well the words spoken on that occasion by my father (Martinaglia 1947a; van Riet Lowe 1947).

Matters moved swiftly after the opening of the caves. We may imagine that the news spread widely and visitors to the newly revealed underground caves were numerous. At the monthly meeting of the Geological Society on 13 September 1897, JE McNellan is reported to have presented to the Society for its museum 'Stalagmite and photos of Sterkfontein Cave and other scenes'. This is the first record of any items having been presented to the museum from the Bloubank River Valley area; it would be valuable to know if those items are still in existence in the Geological Museum of

Fig. 13.4 *Plaque erected at Sterkfontein in honour of Martinaglia's discovery of the underground caves*

Johannesburg. The photograph may be the earliest pictorial record available of the interior of the caves.

At the same meeting, Mr Minett E Frames referred to 'the animal remains found in the Kromdraai Caves in the Dolomite near Krugersdorp'. Among these, he cited remains of 'the horse species, antelopes, monkeys, porcupines, rats, bats, etc.'. And he added:

...the presence of the two first in the cave would lead one to infer that they had been dragged there by beasts of prey. The cave is now filled with bones, cave earth, and stalagmites to a depth of about 15 feet, and the whole mass is consolidated into hard rock. The latter feature naturally points to the antiquity of the cave and its contents. The roof has, in some cases, been eroded, so that portions of the bone bed are today exposed on the surface (Frames 1897: 95).

This remarkable statement anticipates by sixty to seventy years Brain's hypothesis that the major part of the bone accumulations in caves is due to the activities of leopards or sabre-toothed cats. Here Frames appears to have been referring to the Kromdraai caves. Frames comments, too, on the absence of 'Arctic species' from the deposit and the presence of 'living types', and states that the evidence from here and elsewhere in South Africa points to a 'persistently warm climate'.

All the evidence we have adduced so far shows that a rich fossil site was known and had been reported at Kromdraai since 1895; none had yet been reported at Sterkfontein, and even Draper's statement to the Geological Society on 12 July 1897 had not mentioned fossil bones from Sterkfontein.

On the other hand, there is little doubt that the caves referred to by Dr Hugh Exton, MD, FGS, in his presidential address to the Geological Society on 22 February 1898, were those known today as the Sterkfontein caves. He refers to the 'opening out of the stalactite caves at Sterkfontein' and stresses the remarkable beauty of 'the branching crystals standing out from the pendant stalactites, the transparency and brilliancy of the crystals reflecting the light of the lamps held by visitors...' He stated that these crystals were of a form of calcium carbonate known as aragonite (Exton 1898: 7–8). It is noteworthy that, despite Exton's devoting a long paragraph to the Sterkfontein caves, he makes no mention of fossil bones (and Exton was a medical man, who had at one time been district surgeon in Bloemfontein!).

Thus, neither Draper (1897) nor Exton (1898), in speaking of the Sterkfontein caves, had made mention of fossil bones. Is it possible that none was known from these caves until 1898?

Probably in response to a report in the English journal *Mechanic* of 27 August 1897, which was republished in the French periodical *Cosmos*, members of the Marist Brothers School in Johannesburg undertook a visit to the Sterkfontein caves.

It is most likely that the Brothers would have planned their visit during the school holidays, so the visit probably took place in January 1898. Eight people went out in a carriage or wagon, taking four hours to get to the caves. There they found Mr Thomasset in charge (Malan later tracked the history of Thomasset; see Malan 1959: 322–323).

Fig. 13.5 *Five Marist Brothers with Hans Paul Thomasset at the entrance to the Sterkfontein caves, about January 1898*

Thomasset went down alone into the caves, lit a fair number of candles and when all was ready came back to the surface to fetch the Brothers. One of the Brothers described how they entered the caves, as follows:

> First we go down to the bottom of the excavation which was made before the discovery of the caves to extract limestone, and we find ourselves at the entrance to the caves (I am sending you the photograph of this taken by ourselves).
>
> This entrance is very large; you would think it was the hall of a mansion. You go to the right and after five or six metres you must go down vertically for about ten metres by means of a ladder and you find yourself in the biggest hall which seems to be about 20 metres high and 30 metres wide. It is impossible to describe the sensation that you feel on seeing the works of nature in this cave. You would think it was of crystal. Thousands of stalactites of different shapes and sizes hang above one's head. Several have long since become joined to stalagmites and form magnificent pillars. In certain places they resemble organ pipes. Mr Thomasset started to strike them with a stick when we weren't expecting it; we thought we were hearing real bells (Un Frère Mariste 1898: 133–135).

The published description appeared in *Cosmos* (Paris) in 1898 and dwells on the size of the caverns, the beauty of the crystalline formations, the stalactites and stalagmites, several of which had become united to form continuous pillars. In free translation, the expedition's scribe wrote:

> You cannot see any rocks there; everything is covered with a sort of vegetation of crystals of infinite variety. Here, you would say, they were mushrooms, there cauliflowers, further away ostrich feathers, moss, and everywhere it seems as if needles, pins, pearls and diamonds have been strewn pell-mell in great profusion; but everything is so firmly joined together and so hard that you need a hammer to take samples.

Then came the crucial find, perhaps only one year or less since the caves had been blasted into by Guglielmo Martinaglia. The scribe writes:

> In a certain place, a metre from the ground, I found animal fossils, a jaw complete with all its teeth. It must have been the jaw of a large antelope, probably the victim of carnivorous animals.

This article is the first definitive account of the Sterkfontein caves and the first published account of the presence of fossil bones in them. From an early stage the priceless treasures of Sterkfontein were subject to vandalism and souvenir seeking. In a footnote to an article published by Professor GAF Molengraaf (appointed state geologist in the South African Republic on 7 September 1897) the author writes:

> Many of these grottoes are very beautiful and worth visiting. One discovered in 1897 at Sterkfontein (68), to the north of Krugersdorp, was a splendid natural curiosity; stalactitic draperies adorned the walls, and the stalactites were entirely covered with magnificent needle-shaped crystals of arragonite [sic]. The numerous visitors who arrived from Johannesburg destroyed this marvel in a few weeks' time, and as the grotto was on private property the government unfortunately could not interfere (Molengraaf 1904: 31).

Much later, in 1920, HG Nolan, who had earlier partnered Martinaglia, is said to have completed this devastation in a fit of pique after his lease was not renewed. According to Martinaglia, Nolan told his late brother that they had 'stuffed the caves with dynamite and destroyed nearly all the stalactites and stalagmites' (M Nolan in van Riet Lowe 1947).

The Sterkfontein caves after the turn of the century

Fascinating information about the Sterkfontein caves has been gleaned and recorded by Mr Harry Zeederberg, sometime secretary of the West Rand Historical Society. He learnt much from the old pioneers of the district in the decade from 1950 to 1960.

During the South African War, according to Zeederberg, when Krugersdorp was occupied by British troops, the caves were used as an important food and ammunition depot. A detachment of the Scots Guards, under Sergeant-Major Tommy Llewellyn, lived on the hillside for nearly a year and guarded the depot in the caves.

Zeederberg has long maintained that the Sterkfontein caves were discovered before the Jameson Raid and not afterwards. This is based on the recollections of such pioneers as PW Willis, Andries te Water and Frans Dekker. Some of these claims are to the effect that lime was being worked at Sterkfontein before the Raid. On this we can, I think, agree: there seems little doubt that lime working was going on from about 1893. When the Marist Brothers paid their visit, they recounted that they had to go down into a quarry – 'First we go down to the bottom of the excavation which was made before the discovery of the caves to extract limestone.'

Interest in the fossil possibilities of South Africa's caves was spectacularly aroused after 31-year-old Professor RA Dart's revelation of the Taung skull in 1924 and his publication of it in 1925 (Dart 1925). It was a representative of a new and unprecedented type of higher primate to which Dart gave the name of *Australopithecus africanus* (southern ape of Africa).

Fig. 13.6 Mr Robert M Cooper, mayor of Krugersdorp 1925–26, and sometime holder of mineral rights at Sterkfontein

Against this background, there came to the Sterkfontein caves in 1929 a remarkable man, Robert M Cooper, a town councillor and former mayor of Krugersdorp (Fig. 13.6). In that year he acquired mineral rights of the property.

It has been commonly believed that, between those early years of the late 1890s and 1935 when Trevor Jones (1937) collected fossil baboons from Sterkfontein, no scientist took any interest in the Sterkfontein caves. Thus Broom (1946: 46) states, 'It is sad to think that for nearly 40 years no scientist ever paid the slightest attention to these caves,' and again, 'Trevor Jones who visited the caves in 1935 appears to have been the first scientist who was interested in them for 38 years.' My investigations have revealed that this was not entirely correct.

In spite of the belief that no scientist took an interest in the bones from Sterkfontein, not long after Dart (1925) had published an account of the 'Taung child' that would ultimately move the spotlight from Asia to Africa, he received a box of bone-bearing breccia from Sterkfontein. As Dart and Craig put it:

In South Africa miners in lime quarries were now on the lookout for bones, and boxes of breccia were turning up frequently at the Medical School for my inspection. Among these were lime-consolidated bones from Sterkfontein... The Sterkfontein bones included the skull of a large baboon not greatly dissimilar from the living form, so I concluded that geologically this deposit must be relatively recent compared with Taungs... 1959: 50).

Interestingly, and with a nice historical irony, the most recent researches have shown that the relative dates of the fossils from the two sites were exactly the other way around – Taung proving, in the event, to be younger than Sterkfontein.

This box of bones received from Sterkfontein, because of its presumed young age, seems to have suffered the same fate as that which attended the bone breccia sent to the British Museum in 1896.

Again in 1931 or 1932 Professor JHS Gear, who had been one of Dart's former research students, visited Sterkfontein with two Krugersdorp friends, Carl and Piet Goedvolk. In a letter to me (Fig. 13.7), Professor Gear wrote on 20 July 1972:

Dear Professor Tobias,

Many thanks for your letter of 13th July. I have just spoken by phone to Dr. Carl Goedvolk, who says that his memory of our visit to the caves is now somewhat hazy. However, he lived in Krugersdorp and he and his brother, as boys, had explored the caves and knew of the presence of fossil bones. In 1931 - 32 after my appointment to the Institute, I was writing the account of the "Fossil Baboons from Taungs" published in the Dart Festschrift number of the Leech many years later, and so knowing of my interest in fossils Dr. Goedvolk arranged that he and his brother, who was then a school teacher, and I should explore the caves, which at that time had not been mined for limestone.

We went one Saturday in 1931 or 1932, probably 1932 and spent a large part of the day underground collecting bones, filling three to four wooden boxes, of the kind used at the Institute at that time to pack vaccines and sera. Most of the bones were of buck, but there were several fragmented baboon skulls, but we did not find the remains of any Taung's type "ape-men", which was the main purpose of our search. I noted that there were several "leads" from openings on the surface down to the bottom of the caves, through which water was dripping and would flow following rain. At the top of these leads the soil was earthen but at the bottom it was largely limestone. There were bones embedded in the earthen soil as well as the limestone from the surface to the bottom of the cave. We did not explore the site where we met you and your party at Sterkfontein, and most of our specimens were obtained from within the caves which had not then been quarried for limestone.

We brought the boxes back to the Institute hoping to study the bones. However, due to the pressure of other work including taking the course for the D.P.H. immediately after which I left for London, this was not done in detail.

In the course of conversation usually at the Medical School lunch table, I mentioned our findings to various members of the Anatomy Department and suggested the caves would be well worth exploring further.

I have asked our store to look for the boxes, but it is feared that they were discarded. However, they are looking for them and if they can be found, I would be delighted to hand them over to you.

Many thanks for your kindness in telling our American guest, Dr. Patricia Larson of the work which had been done at Sterkfontein. We appreciated it very much and so did she.

Many thanks too for your congratulations on the award of the Fellowship of the Royal College of Physicians.

With best wishes,

Yours sincerely,

J.H.S. GEAR.
DIRECTOR.

Fig. 13.7 Letter from JHS Gear about his recovery of fossils from Sterkfontein in 1931 or 1932

In 1931–32 after my appointment to the Institute, I was writing the account of the "Fossil Baboons from Taungs" published in the Dart Festschrift number of the *Leech* many years later, and so knowing of my interest in fossils Dr Goedvolk arranged that he and his brother, who was then a school teacher, and I should explore the caves, which at that time had not been mined for limestone [sic].

We went one Saturday in 1931 or 1932, probably 1932 and spent a large part of the day underground collecting bones, filling three to four wooden boxes, of the kind used at the Institute at that time to pack vaccines and sera. Most of the bones were of buck, but there were several fragmented baboon skulls, but we did not find the remains of any Taung's type 'ape-men', which was the main purpose of our search.

A search at the South African Institute for Medical Research in 1972 failed to reveal these bones. This is a great pity, because, if they were obtained from the exposed lowest breccia inside the *underground* cave system, they might well have been the oldest bones so far recovered by a scientist from the Sterkfontein breccial deposit.

Some time towards the middle of 1933, Dr Giovanni Martinaglia collected insect specimens in the Sterkfontein caves and submitted them to the Imperial Institute for Entomology for identification. On 21 June 1933, that Institute reported that the dipteran flies submitted belonged to two species of Streblidae: of one, *Nycteribosca africana*, Walk, there were only two specimens previously in the British Museum; the other, *Raymondia planiceps*, Jobling, was new to the British Museum.

About 1935, Trevor Jones managed to obtain some fossil baboons from Mr RM Cooper; these specimens had come from the Sterkfontein cave. In *Beyond Antiquity* (Dart 1965), Jones describes how he acquired his first three specimens from Mr Cooper's shop in Krugersdorp: in the window was an advertisement, 'Buy your bat guano from the Sterkfontein caves and see the Missing Link'. On a subsequent visit to the Sterkfontein caves Jones states that he was allowed by the caretaker (Mr GW Barlow, who had previously worked at Taung) to take away about forty specimens (Dart 1965: 61ff.). In 1936, Jones developed and studied these baboon skulls as part of his BSc honours work in the Wits University Anatomy Department. He subsequently presented a paper to the Jubilee Congress of the South African Association for the Advancement of Science in Johannesburg in October 1936 (Jones 1937). Among the fossil baboons, he recognised one new species which he named in honour of Dr Broom, *Parapapio broomi*.

Harding le Riche was the next link in the chain. Armed with a baboon endocast that Mr Barlow had given him at Sterkfontein, he aroused the interest of Schepers. Apparently independently of Trevor Jones, these two other members of Dart's department went out to Sterkfontein in 1936. Gerrit WH Schepers was a final-year medical student and graduate demonstrator on Dart's staff, and Harding le Riche was a third-year medical BSc student. Le Riche was afterwards to become Professor of Epidemiology and Biometrics at the University of Toronto. Schepers later had the privilege of describing the endocranial casts of the australopithecines and was thus a co-author with R Broom of two of the notable memoirs published by the Transvaal Museum, Pretoria.

After having worked through the previous night until 4 a.m., they took some of their fossils across to Dr Broom on Monday afternoon, 3 August, and gave some of the material to him. According to Findlay's account, and Wells's memory, they probably did this without the knowledge of Dart and some unpleasantness seems to have ensued (Findlay 1972: 66).

Whatever the rights or wrongs of the matter were, there is no doubt in my mind from a careful study of the available statements, as well as personal discussions, that Broom was prompted to go to Sterkfontein, not by Jones's pelvis, but by le Riche's and Schepers's baboon skulls and brain-casts. These two first took Broom to the Sterkfontein caves. The stage was set for the Broom era and the first period of hominid discoveries.

The Broom era: The first period of hominid discoveries, 1936–39

The first eight days of Broom's visitations to Sterkfontein were a time of hope with a phenomenal outcome. On Sunday 9 August 1936, in company with le Riche and Schepers and in the motor car belonging to Schepers's uncle, Broom paid his first visit to Sterkfontein. There he met GW Barlow, the manager, and learnt that Barlow had previously worked at Taung. What is more, he had been there at the time when M de Bruyn blasted out the Taung skull in about October 1924. Some fossil animals were recovered from Sterkfontein on that first visit by Broom.

The second visit by the 70-year-old Broom took place three days later on 12 August, when Broom received from Barlow 'three nice little fossil baboon skulls' and much of the skull of a large sabre-toothed cat (Broom 1946: 46).

The third visit was on the eighth day after Broom's first visit: 17 August. Barlow handed over to Broom the blasted-out natural brain-cast of a hominid and Broom said, 'That's what I'm after' (Broom 1950: 44). This remarkable good fortune, to find a missing link eight days after beginning his search at Sterkfontein, characterised much of Broom's scientific career and led him to believe that spiritual forces in nature were guiding him to the right spot at the right time.

In this way the first authenticated skull of an adult australopithecine came into Broom's hands. He at first called it *Australopithecus transvaalensis*, that is, of the same genus as the 'Taung child', but of a different species. When further specimens were recovered by Broom later, he judged that the Sterkfontein ape-man was so different from that of Taung as to justify his placing it in a separate genus, for which he invented the name *Plesianthropus*; *plesios* means 'near' and *anthropus* means 'man'. Whereas Dart had modestly called his creature *Australopithecus,* the 'southern ape', Broom thereby boldly proclaimed his belief that the specimen from Sterkfontein was close to man – *Plesianthropus*. By 1954, many more specimens were available and detailed studies had been made. It was then possible for John Talbot Robinson to propose that the Sterkfontein fossils should be reassigned to not only the same genus, but the same species as the 'Taung child', though to a different sub-species or race, which he called appropriately *Australopithecus africanus transvaalensis*.

The discoveries in these Transvaal sites had made little impact on anthropology by 1939. The world did not seem to know what to make of the flood of short communications that Broom sent to *Nature*, characterised as they were by hasty compilation, freehand (albeit inimitable) drawings, the absence of comparative data or of references, and somewhat cavalier taxonomic procedures. Another reason why the fossils did not make the impact they should have done was Broom's judgement that Sterkfontein was of relatively recent geological age. He regarded Taung and even Kromdraai as older by far than Sterkfontein, which he set at first in the Upper Pleistocene (Broom 1938a) and subsequently in the Middle Pleistocene (Broom 1938b).

This supposedly youngest age of Sterkfontein led Broom to assign lesser importance to the fossils from that site, for the younger the fossils were, the less the likelihood that they were ancestral to humans. So in Broom's eyes, the importance of Sterkfontein was eclipsed by that of the putatively somewhat earlier site Kromdraai, and both of these by Taung, assumed to be the oldest of the three. It is interesting to reflect that recent research has suggested that the relative dating of the three sites was probably the reverse of what Broom at that time believed, with the deepest layers of Sterkfontein the oldest and Kromdraai perhaps the youngest!

Dr HK Silberberg (Fig. 13.8) was a well-known Johannesburg art dealer and collector as well as an antiquary. He was deeply interested in proto-history and prehistory. From 1936, when Broom made his first australopithecine find at Sterkfontein, Dr Silberberg visited and collected specimens in the deeper part of the Sterkfontein caves.

Fig. 13.8 Dr Helmut Kurt Silberberg, after whom the Silberberg Grotto at Sterkfontein is named

Fig. 13.9 Dr HK Silberberg's sketch of the fossil hyena locality deep in the Sterkfontein caves (from his letter to PV Tobias)

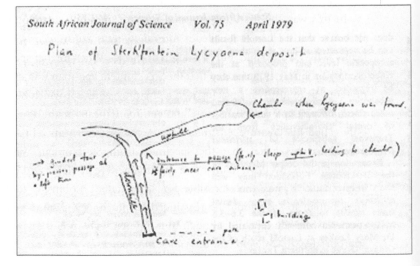

He collected only pieces that he found on the floor of what Murray Justin Wilkinson (1973) proposed should be called the Daylight Chamber (Fig. 13.9). To this chamber the University of the Witwatersrand late in 1978 gave the name of Silberberg Grotto for reasons which I am about to unfold.

Between 1942 and 1944, probably 1942 (Tobias 1979), Dr Silberberg collected from this grotto pieces of bone breccia containing baboon and hyena remains. For some time he kept the handful of specimens of breccia in his gallery in Johannesburg. In January 1945, he showed them to the Abbé Henri Breuil, who 'seemed fascinated by the hyena fossil' (Silberberg 1979). Dr Silberberg gave the fossil to the Abbé who, in turn, took it and some others across to Broom in Pretoria. When Broom addressed another of his epistles to *Nature*, on 27 January 1945, he stated, 'I have just had given to me a few days ago by the Abbé Breuil the snout of a primitive hyena.' Broom had immediately realised the great significance of the hyena find and within days, as was his wont, he had penned a letter to the editor of *Nature*. Broom identified the hyena as an extremely primitive form akin to *Lycyaena*, a Lower Pliocene genus found in Europe and India. Broom declared, '…the evidence it affords seems to be in favour of the deposit belonging to some part of the Pliocene rather than the Pleistocene…It thus seems probable that the Sterkfontein cave deposit with its sabre-tooths and its Lycyaena and its absence of horse remains will prove to be Upper Pliocene, and the Kromdraai deposit to be Lower Pleistocene' (Broom & Schepers 1946).

Thus, the discovery of Silberberg's hyena was directly responsible for Broom's change of mind about the relative ages of Sterkfontein and Kromdraai. He now made Sterkfontein the more ancient. Moreover, he now felt constrained to remove Sterkfontein from the Pleistocene and place it in the Pliocene (where all later research would place the Sterkfontein stratum from which *Australopithecus africanus* was derived).

Only in 1946 did Broom systematically describe and analyse the Cradle's hominid fossils, in the first of the volumes he published through the Transvaal Museum. Only now was a hitherto sceptical, even disbelieving world of science persuaded of the significance of the finds. Many hundreds of ape-man fossils that have been laid bare at Sterkfontein, Swartkrans and Drimolen since the 1970s have given the world its two richest cave sites for hominid remains and have done much to establish the role that Africa played in human evolution. Finally the Cradle and the scientists who had worked there received the recognition they deserved, when the fossil hominid sites of Sterkfontein, Kromdraai, Swartkrans and the surrounding land (collectively given the brand name, 'Cradle of Humankind') were listed by UNESCO as a World Heritage Site on 2 December 1999.

CHAPTER 14

THE SOUTH AFRICAN WAR OF 1899–1902 IN THE 'CRADLE OF HUMANKIND'

Vincent Carruthers

On 10 October 1899 war was declared between the British government and the South African Republic and its ally, the Orange Free State. The British believed the war would be over by Christmas, but it was to drag on for two-and-a-half years during which fighting spread deep into rural South Africa, and by 1900 the Bankenveld – the rough mountainous country between the Magaliesberg and Krugersdorp, including the 'Cradle of Humankind' – had become one of the principal theatres of war.

The importance of the area can be attributed partly to the leadership of the Boer general JH 'Koos' de la Rey, who repeatedly thwarted British dominance west of Pretoria and who was probably the most skilled of all the generals on either side of this war. A political opponent of President Paul Kruger, he had argued strongly against Boer entry into the war. Shortly before the outbreak of hostilities Kruger charged him with cowardice in the Volksraad. De la Rey's retort to the president was powerfully prophetic: 'I shall do my duty as the Raad decides, and you, you will see me in the field fighting for our independence long after you and your party who make war with your mouths have fled the country' (B Fairwell quoted in Cloete 2000: 29). Within a year Kruger had been shepherded into ignominious exile while de la Rey's remarkable success in the field continued until the very end of the war. His iconic prestige among friend and foe gave him enormous stature at the peace negotiations in Vereeniging and considerably influenced their outcome, and hence the future of South Africa.

A second reason for the military importance of the Bankenveld was its topography. For the British the mountains offered a natural barrier against which to drive the elusive Boer guerrillas, while for the Boers, mountainous and complex terrain provided the ideal situation in which their small, veld-wise commandos could ambush and harass

*Fig. 14.1 General JH 'Koos' de la Rey.
Drawing by Anton van Wouw*

their enemy. As guerrilla warfare replaced conventional field battles, the British retaliated by escalating their scorched-earth policy to include widespread civilian involvement. The region became the testing ground for these fundamental shifts in the nature of the conflict, and a bitter resentment was born that would imprint itself on subsequent generations and influence South African politics for the next century. The Cradle region provides a lens through which to view this significant moment in South African history.

At the outbreak of war the Boers successfully followed the military plan that had been proposed by Jan Smuts, who was then the attorney-general of the Transvaal: 'The Republics must get the better of the English troops from the start…by taking the offensive, and doing it before the British force now in South Africa is markedly strengthened' (Hancock 1962: 111). Within a few weeks they had driven the British back into the Cape and Natal colonies and immobilised them in a series of sieges. By February the following year, however, the Boers were in full retreat. Lord Roberts, an ex-Indian army General in his late sixties, had been recalled from retirement to lead the British forces after the initial setbacks. At the same time reinforcements flooded into South Africa from all over the Empire and when Roberts took command in Cape Town his army outnumbered the entire white population of the South African Republic. Such a force was unstoppable, and month by month the Boers were driven back. Most of the commandos surrendered their arms and returned dejected to their farms.

The region now known as the Cradle of Humankind World Heritage Site fell within the Krugersdorp District and the farmers of the area joined the Krugersdorp commando. Their leader was Sarel Oosthuizen, a large, powerfully built red-haired man with the nickname 'Rooibul' (Red Bull). This commando was among those who invaded the colony of Natal at the outbreak of war and contributed to the early Boer victories of 1899 and the siege of Ladysmith. But the battlefields of Natal were a long way from home, and the Krugersdorp men probably never suspected that the fury of war would ever reach their own quiet highveld farms.

The battle of Kalkheuwel, 3 June 1900

The peace of the Bankenveld valleys was shattered in early June 1900. Roberts was ready to begin his triumphal march to Pretoria, President Paul Kruger's capital. With military bands playing and colours flying the main body of the army moved directly along the road from already-occupied Johannesburg, while two powerful cavalry brigades, each of about 4 500 horsemen and several field guns, encircled Pretoria from the west. One of these columns was led by Lieutenant-General Sir John French, the most senior cavalry officer in the British army. (Later in his career he was to lead the British army into France during the First World War.) His route followed what is today the R512 road past

Lanseria Airport. After crossing the Crocodile River the road twists through the Kalkheuwel hills, a high ridge honeycombed with lime mines and the dolomitic caves that make the World Heritage Site famous. A century ago this was a dangerous, narrow defile flanked by rocky slopes that confined the mounted troops to the track and were difficult to climb, even on foot. It was here that a commando led by Sarel du Toit lay in ambush. Their intention was to hold the British at bay for long enough to allow a Boer wagon train loaded with supplies and refugees to elude capture and slip away safely to the north.

The cavalry advanced slowly, encountering small parties of Boers. most of whom surrendered their arms and signed an oath of neutrality. By the time they entered the Kalkheuwel pass, long winter afternoon shadows darkened the wooded slopes and little more than an hour or two of daylight remained. As the vanguard reached the midpoint of the pass du Toit's men, concealed among the rocks and shadows, opened fire at almost point-blank range. Two British soldiers were killed immediately and another was mortally wounded. The leading troops tried to gallop back but were blocked in the narrow road by others pressing forward. For a while there was pandemonium until a troop of New South Wales mounted infantry stood their ground, and their example restored order to the British column.

Shells and rifle fire poured into the valley. Dragoons dismounted and took cover behind their wounded horses. French brought up his artillery and soon '…the noise was deafening: guns, pom-poms, Mauser and our own fire – one continuous roar, echoed and re-echoed among the hills' (Morton, *Diary of the 8th Hussars*, quoted in Copley 1993: 118).

Gradually the British regained the advantage, but Boer sniping and counter-fire continued and kept them pinned down throughout the night. Just before dawn the Boers retreated, leaving more than twenty men dead and several abandoned wagons. Losses on the British side had been remarkably light, although many horses had been killed.

The fall of Pretoria and Boer reorganisation

Lord Roberts had expected the fall of Pretoria to end the war. The Union Jack flew above his victory parade in Church Square but the enemy was not yet vanquished. The remnants of the Boer army had evacuated the capital without defending it and, under a younger and more vigorous leadership, had resolved to fight on. The new strategy was to deploy four autonomous regional commanders throughout the Transvaal and to use guerrilla tactics to wear the enemy down. General de la Rey took command of the western Transvaal. Under him were several field generals, one of whom was 'Rooibul' Oosthuizen of the Krugersdorp commando. The responsibilities of the regional commanders were to rekindle the fighting morale of the farmers, re-recruit them into the

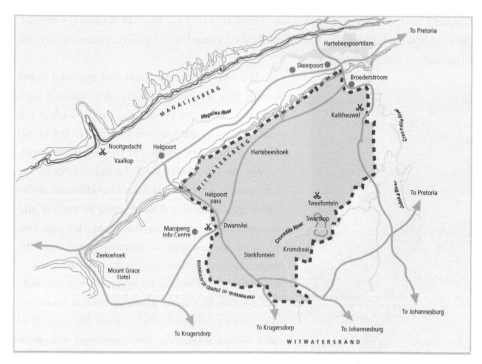

Fig. 14.2 A map of the main South African War actions in and around the area that would later be known as the 'Cradle of Humankind'.

commandos and lead opportunistic attacks against the British. The new commandos comprised smaller units – about one hundred men under a veld-cornet – and they had to be self-sufficient for their day-to-day supplies. Wherever possible, they were based in their home districts where they were familiar with the territory and close to their farms and families. Discipline was tightened and casual leave was no longer allowed. Veld-cornets and commandants were appointed by their superiors, not elected by their men as had previously been the case.

Oosthuizen returned to revitalise resistance in the Krugersdorp–Magaliesberg region. Re-recruiting was not always easy among the dispirited burghers who had quit the war and could see no merit in a futile struggle against what they believed were over-whelming odds. Many had taken an oath of neutrality, and there was bitter dissension between *hensoppers* (literally 'hands-uppers', Boers who continued to honour the oath) and the *bittereinders* ('bitter-enders') who were prepared to continue the fight. The majority, however, had taken the oath reluctantly and the arms surrendered to the British were frequently antiquated hunting rifles, while the modern, military-issue Mausers were hidden in caves or buried along with other family treasures such as crockery and silverware. Oosthuizen seems also to have been extremely persuasive, because in a matter of weeks he had established a number of small commandos between Krugersdorp and Hekpoort.

Fig. 14.3 Max Weber, who served with and recorded the actions of the Swartkop commando.

The Swartkop commando that operated in the Cradle area was one of Oosthuizen's units. It took its name from the prominent, saddle-backed hill that dominates the south-eastern part of the World Heritage Site. Frans van Zyl was appointed veld-cornet, and he established the base camp of the commando on his own farm 'Tweefontein', today called the Cradle Game Reserve and Restaurant and situated in the centre of the Word Heritage Site. Most of the commando members were from the immediate neighbourhood and by day they stayed on their own farms, carrying out their commando duties only after dark. Much of what is known of this small commando comes from the journal of Max Weber, a German–Swiss geologist, who published his account in a Zurich newspaper after the war. Weber's diary is a first-hand record of life in the Swartkop commando, much of it told with a detached, even critical, perspective on his comrades-in-arms. Weber had joined the Boer forces at the outbreak of war and after being wounded in battle on the Thukela River, he returned to his home on the Jukskei River just north of Johannesburg. But in the spring of 1900 he joined the Swartkop commando. Some two thousand foreigners actively supported the Boer forces during the war but, as Weber records, their assistance was seldom welcomed with enthusiasm (Weber n.d.).

The Swartkop hill rises 200 metres above the Crocodile River and the summit commands a magnificent view, from Krugersdorp to Johannesburg in the south and to the Magaliesberg in the north. From here the commando could observe and report on the movement of British soldiers around Krugersdorp, and conduct occasional raids to replenish supplies and disrupt the enemy. The information they gathered was critical to Oosthuizen's plans and almost certainly contributed to the success of his first assault.

The battle of Dwarsvlei, 11 July 1900

While Oosthuizen was rallying the commandos around Krugersdorp, a fellow field-general, HL Lemmer, was enjoying similar success around Rustenburg. Soon the entire

Fig. 14.4 The distinctive Swartkop hill from which the local commando took its name.

Bankenveld was in Boer hands and the British garrison in Rustenburg was isolated from the British high command in Pretoria. To those living in the district it was evident that the war was far from over. Sarah Heckford, whose extraordinary career had varied from founding a famous London children's hospital to itinerant trader in the far northern Transvaal, was at that time employed as governess in the pro-British Jennings household on the farm 'Nooitgedacht' near Hekpoort. As Transvaal burghers, her male employers had been recruited into the Boer commandos at the outbreak of war, but their sympathies were respected, and they had been allowed to hold non-combatant positions. But with the commencement of the guerrilla war the attitudes of the Boer command were hardening, and non-combatants were a luxury they could not afford. Consequently, the men from 'Nooitgedacht' made their escape from the farm and went over to the British side. Realising that the commandos were reviving and gathering, Heckford rode alone into Pretoria to warn Lord Roberts. Similar intelligence was reaching him from other quarters, and Major-General HA Smith-Dorrien was sent to deal with the situation.

On 11 July Smith-Dorrien led 1 300 Gordon Highlanders and other infantrymen and horse artillery out of Krugersdorp to force the reopening of the road to Rustenburg and to escort a convoy of 40 supply wagons to that beleaguered town. The plan was to follow the main road from Krugersdorp to Hekpoort, where they were to be joined by a cavalry regiment of Scots Greys. By then, Scottish regiments had taken to wearing khaki aprons over their plaid kilts as camouflage in the winter veld, but their departure was known to the Boers. Certainly the lookouts on Swartkop would easily have monitored the movement of such a large column. There would have been considerable excitement among the Boers, for this was to be the first encounter fought by General Oosthuizen's newly recruited and enthusiastic commandos.

The battlefield Oosthuizen chose was his own farm, 'Dwarsvlei', close to Maropeng, the modern Cradle information centre. It was an ideal position for an ambush. The Witwatersberg ridge overlooks the Krugersdorp–Hekpoort road at the crossroads that today leads to the Hartebeesthoek Tracking Station in the west and to Maropeng in the east. Oosthuizen did not have a large force – some 750 men to confront a column of twice that number – but he deployed them along the crests of the surrounding hillocks and they waited in silence as the British marched up the hill. When the column reached the intersection, the Boers opened fire.

The battle that followed was distinguished by extraordinary acts of courage on both sides. Smith-Dorrien's artillery galloped forward into a gap near the top of the ridge, but Boer riflemen held the high ground on either side of them. Within half an hour fourteen of the seventeen gunners had been hit; their horses were shot down in their harnesses. Wounded men lay trapped with dead colleagues behind hastily constructed earth and stone defences. The section commander, Lieutenant Turner, who was wounded three times, continued to try to fire the guns alone. Later, Captains WE Gordon and DR Younger of the Gordon Highlanders each won Victoria Crosses by leading an attempt to manhandle the guns to safety. Younger and three other Highlanders were killed in the endeavour and a fourth man, Corporal JF Mackay, was recommended for the award for carrying Younger's body out of the firing line.

Despite the disparity in combatant numbers, the battle continued all day with the Gordon Highlanders holding off heroic Boer attempts to capture the guns and wagons. At dusk General Oosthuizen personally led a final and desperate charge against the guns. Just a few metres from his objective he was mortally wounded, and died some weeks later. With the loss of their leader the Boers fell back and once it was dark, the British managed to limp back to Krugersdorp.

In an irony that often attends the heritage of war, the graves of Sarel Oosthuizen and Captain Younger lie a few metres apart in the Krugersdorp cemetery. No other monument commemorates that day. Indeed, the new tarred road leading to Maropeng now obliterates the spot where the Boers lost one of their most gallant generals and where two British soldiers won their country's supreme decoration for valour.

The growth of Boer resistance

Although neither side could claim decisive victory from the battle of Dwarsvlei, the Boers had succeeded in keeping Rustenburg isolated. They were further encouraged by two other successes on the same day. At Silkaatsnek, east of Hartbeespoort Dam, de la Rey had destroyed the cavalry column that was to have joined Smith-Dorrien at Hekpoort, while in a skirmish north of Pretoria, a British outpost had been chased back into the town with the loss of several men.

The British did not recognise the significance of this trio of events at the time. They believed them to be mere setbacks, and were sure that Boer capitulation was still not only inevitable but imminent. For the Boers, however, the events of that day had been extremely significant. They had demonstrated a new-found effectiveness, and many others were emboldened to continue the war. Consequently, re-recruitment now surged ahead. Max Weber expressed the mood: 'Then something sounded through the air! A vague rumour that went from mouth to mouth told of heroic deeds, of war and battle, of a wild free life on horseback. The reports became more definite till at last the facts became known: The war is not yet over! General de la Rey is returning with great commandos' (Weber n.d.: 3).

After Dwarsvlei the Boers were in possession of all the rural areas from Swartkop to the Magaliesberg. For this reason, the role of the Swartkop commando changed from being purely one of observation and intelligence to more active raiding for food and other supplies. Occasionally these attacks were against British patrols in order to obtain

Fig. 14.5 Defence and lookout post on 'Tweefontein'

saddles and arms, but the favourite targets for everyday supplies of food and clothing were the *hensoppers*, regarded as traitors deserving to be robbed so as to sustain their compatriots in the field.

Another function that the commando now took upon itself was to regain what it regarded as 'proper control' of black people in the area. British occupation had introduced legal rights not previously enjoyed by blacks in the South African Republic. Most disturbing to Weber and his compatriots was the right of Africans to give evidence against white farmers, especially with regard to violations of the oath of neutrality. While Africans in the Cradle district were evidently never armed, Weber refers to them as 'spies' and claims that several farmers in the Kromdraai valley were punished and their homesteads destroyed by the British, on the evidence of black strangers visiting the district (Weber n.d.: 71–74). Blacks had also apparently 'become restless and no longer wanted to work on the farms' under the less oppressive British rule. Later Weber records their taking advantage of the situation and actively trading beer for British clothes and the clothes for Boer money (Weber n.d.: 41–42). With the district back under Boer control, the Swartkop commando saw it as their duty to reverse the situation. Unbeknown to Weber and his colleagues, however, the question of justice was a matter of some contention between the general officer commanding the Krugersdorp District, Major-General G Barton, and the quasi-civilian court set up by Colonel C Mackenzie, the military governor of the Witwatersrand. On the matter of evidence, Barton might have found common cause with Weber, because he believed that the magistrate's court 'required evidence out of proportion to the circumstances of war and marshal law'. Mackenzie commented that Barton's complaint was simply that the 'magistrate does not find every man accused of a crime guilty to the degree that General Barton desires him to be' (Spies 1977: 70).

The escapades of the Swartkop commando attracted a cross-section of outlaws and fugitives from the cosmopolitan community in Johannesburg whose sympathy with the Boer cause was often not entirely honourable. One of the more colourful of these characters was an Englishman whom Weber refers to only as 'Bell'. Bell, it appears, had deserted from the Imperial Yeomanry in Johannesburg in order to escape from justice after he had murdered a black man. Among his fellows in the commando he soon acquired the reputation of a Pimpernel, because of his brazen impersonation of an English soldier that enabled him to gain access to and raid British outposts for supplies. In doing so, he served the interests of the commando well while amassing a substantial personal fortune at the same time.

Weber frequently accompanied Bell on his raids and writes amusingly (and no doubt exaggeratedly) of Bell's courtesy when plundering a surrendered British outpost: 'Do you mind handing over your trousers? Here is your watch and purse – no thank you, I am not a robber. It is my men who need the clothes. We have enough money. Allow me to ask for your horse and saddle' (Weber n.d.: 10).

Bell's reputation eventually led to the British putting a price of £1 000 on his head, and a series of strange 'deserters' tried to join the Swartkop commando in order to trap and capture the elusive renegade. None succeeded, however, but their persistence made his presence in the district unsafe and this eccentric product of the guerrilla war was eventually forced to leave the Cradle area permanently.

The scorched earth policy and the battle of Nooitgedacht

Throughout the latter part of 1900 the Swartkop commando raided British outposts on 'Kromdraai', 'Sterkfontein' and other farms in the Cradle area. The continuing inability of the British to prevent these activities, here and elsewhere in the country, was driving them into methods of warfare which would attract international rebuke. Lord Roberts formally approved the 'scorched-earth policy' in September 1900, and columns of mounted infantry scoured the countryside destroying every farm from which the commandos might obtain any form of sustenance or information. In November the policy was extended to include the destruction of all agricultural crops and livestock. On 16 September Roberts attempted (and failed) to demoralise Boer resistance by ordering the removal of the stone cairn at the base of the Paardekraal Monument in Krugersdorp that commemorated the outbreak of the Transvaal War of 1880–1881. The stones were loaded into bags and dumped in the Durban harbour (Spies 1977: 123–124).

In the first week of December General RAP Clements led a farm-burning column into the Cradle area. It camped at 'Sterkfontein' and for the next two days destroyed farmsteads in Kromdraai and the Crocodile River valley. On the third day the soldiers shelled the slopes of Swartkop from gun emplacements on the Krugersdorp ridge, scattering the commandos camped there. In a pincer movement, they tried to trap Frans van Zyl and the remainder of the Swartkop commando at 'Tweefontein', but the commando escaped and rode overnight to safety south of Roodepoort. That this cat and mouse scenario at times bordered on the absurd is apparent from the fact that two days later Clements moved on to Hekpoort, and the Swartkop commando simply returned to their camp and hilltop vigil.

On 8 December Clements and his men set up a large camp on the Jennings farm 'Nooitgedacht', where they were to await reinforcements from Krugersdorp. The main camp was located below a sheer cliff, and some three hundred picket guards and signallers were sent to guard the summit. Being the highest point on the Magaliesberg, it commands a fine view in every direction, and heliographic communication could be maintained with both Krugersdorp and Rustenburg.

Fig. 14.6 Destruction of farms similar to this one was part of the British scorched-earth policy after 1900.

At 'Zeekoehoek' on the opposite side of the valley, near the present Mount Grace Hotel, Clements's movements were being carefully watched by de la Rey and his recently appointed field-general, Jan Smuts. The weakness in Clements's selection of a camp site was obvious to the generals and they called together all the commandos in the district, including the men of Swartkop. By coincidence, General Christiaan Beyers had brought a large commando from the Zoutpansberg and Waterberg into the area, still equipped with their guns and supply wagons, and the combined Boer force outnumbered Clements's column almost two to one.

After midnight on 12 December 1900 Beyers's men climbed the northern slope of the mountain. As dawn broke they drove the pickets and signallers off the cliffs and took possession of the heights directly overlooking the main encampment. From this position they could fire unhindered into the tents below, forcing the British to retreat. Draught animals were shot to prevent them from hauling the artillery, but Clements's men manhandled the guns to safety. De la Rey's commandos then took the camp while Smuts tried – unsuccessfully – to cut off the British withdrawal. While the Boers looted the now-abandoned camp, Clements was able to take up a position on a nearby hill and to pound his own former position with heavy artillery fire. Smuts described the scene as 'pandemonium in which psalm-singing, looting and general hilarity mixed with explosions of bullets and bombs' (Smuts 1994: 150). Either in that pandemonium or perhaps earlier in the battle, Frans van Zyl was killed and the Swartkop commando was left without an effective leader.

General Beyers in the Cradle area

Following the success at 'Nooitgedacht' the Boer forces moved west to a remote farm, 'Naauwpoort', some 30 kilometres south of Rustenburg and beyond British reach. There, on 16 December – a sacred Afrikaner date – they gathered to commemorate the anniversary of the Voortrekker victory over the Zulu in 1838 and the establishment of the independent Boer Republic in 1881 after Majuba. Women and children from neighbouring farms joined the fighting men in reaffirming their loyalty to the Boer cause. Each person present laid a stone on a commemorative cairn, similar to that which Roberts had removed at Paardekraal, and the location was given the name 'Ebenhaezer – stone of our salvation'. The cairn is still there, now surmounted by a masonry monument.

Determined upon vengeance after the ignominy at 'Nooitgedacht', Clements returned from Pretoria to the Hekpoort valley with twice the number of men he had had at the battle. At the same time General French swept into the valley from Krugersdorp. Several Boers were trapped and captured in their pincer movement, including Commandant Ludwig Krause who had been one of the leaders of the 'Nooitgedacht' attack. But most of the commandos managed to escape, and before long the Swartkop commando had returned to its old territory. The base camp was moved from 'Tweefontein' farm to a site closer to the dolomite caves, where rough shelters were erected and perishable supplies could be kept dry in the caves.

General Beyers's very large commando had also escaped the combined attacks of Clements and French. On Christmas Eve this long column of horsemen, guns and wagons moved into the Swartkop area. The locals were enthralled: 'It was a long time since we had seen real Boer cannons close by,' wrote Max Weber. 'In silent admiration we stood by the wayside and greeted the smart artillerymen' (Weber n.d.: 19). Beyers camped at the foot of the Swartkop hill, intending to cross the heavily fortified blockhouse line on the Johannesburg–Pretoria railway and return to the north. The night after Beyers's arrival, the Swartkop commando was called out to help blow up the railway line as well as an armoured train near Roodepoort, in order to create a diversion while Beyers escaped towards the Modderfontein dynamite factory. This was the last significant action by the Swartkop commando in the South African War.

Horse sickness and *boslansers*

Beyers's passage through the Swartkop area had left a devastating legacy: soon after his visit, African horse sickness swept through the valley. This disease, transmitted by a midge of the genus *Culicoides*, is highly contagious and widespread in the northern parts of the country but had not been known on the Witwatersrand before Beyers's arrival. The affected horse suffers respiratory difficulty and mouth frothing, and generally dies within hours of the appearance of these symptoms (see Joubert 1977). The midge is most active at sunset and dawn during the warm, rainy season and, knowing nothing of its vector, the Boers ascribed the equine ailment to poison in the dew. The outbreak of horse sickness had disastrous consequences for the Boer fighting ability in the district, and possibly contributed more to their eventual capitulation than did military contest. Boers seldom went into action except on horseback or with their horses close at hand. It was this mobility that enabled them to fight so successfully against the overwhelming enemy numbers. But within weeks horse sickness had reduced the mounted men in the Swartkop commando to a third of their original strength. Those who lost their horses became footsloggers and a liability to the others. Unable either to fight or to plunder, they nonetheless still required sustenance and the burden of providing for them fell ever more heavily on the remaining men with mounts. 'Eating, cooking, catching vermin, washing clothes, sewing rags and patches together with threads of sinew, these were the tasks of the pitiable, yet invariably cheerful footsloggers' (Weber n.d.: 91). More seriously, the situation also led to leadership squabbles and difficulties in maintaining discipline as the men switched their allegiance from those leaders with military skills to those who were best at plundering (Weber n.d.: 53).

Farm burning compounded these difficulties, as destitute Boers who had hitherto worked as sharecroppers or *bywoners* on the farms of wealthier countrymen faced the choice between being sent into concentration camps or joining the commandos, albeit without horses. The enormous stigma attached to surrender made many chose the latter. Commandos that enjoyed success in obtaining booty were the most attractive to join, exacerbating the difficulties of the dwindling numbers of mounted men. Men without horses were often treated contemptuously and referred to by disparaging names such as *boslansers*, meaning country bumpkins (or worse), and this resentment was directed at even the keenest fighters if for any reason they lost their horses.

As the situation deteriorated some of the men in the Cradle area drifted off to join General Jan Kemp's commando in the west, as he was reported to have horses to spare. But it seems that his reputation had been exaggerated, and for months the horseless Swartkoppers roamed aimlessly about in wagon trains or on foot, desperately waiting to be issued with horses and vulnerable to Kitchener's flying columns of mounted infantry.

Women, non-combatants and concentration camps

From September 1900 the British combined their scorched-earth policy with what soon became known as 'concentration camps'. Non-combatant refugees from farms that had been destroyed were brought into centralised tented camps. By the end of

Fig. 14.7 Women and children evading capture after the destruction of their farms

that year the camps were experiencing appalling sanitary conditions, disease and many deaths. Black farm labourers who were similarly dispossessed were taken to separate, even more rudimentary, camps. Owing to their fear of camp conditions, and perhaps also because of their pride, many Boer women preferred to roam the veld as nomads rather than being incarcerated. In wagons piled high with salvaged belongings, groups of women, children and occasionally elderly men, tried to keep pace with the mobile commandos or to escape to areas less affected by the war (see Spies 1977 for details about these camps).

Krause's capture (mentioned earlier) had been partly the result of his being delayed at a farmhouse in the Cradle area. He had stopped to help a group of women load their possessions onto ox-wagons in anticipation of their farm being burnt the following day. His anger at their plight remained with him throughout his years as a prisoner of war and coloured much of his subsequent career (see Taitz 1996: 135). In June 1901 two concentration camps were established close to Krugersdorp, one for Boers and another for black refugees. Within a few months these were overcrowded and riddled with disease, and every month one in twenty children died. Later in the war the situation in the camps improved slightly, largely through the campaign led by Emily Hobhouse.

Eventual British control of the Cradle area

Despite the farm-burning, horse sickness and the threat of imprisonment in concentration camps, pockets of Boer resistance in the areas north of Krugersdorp persisted into the new year. Indeed, Barton frequently complained that anti-British sentiment was particularly strong in this area. In April 1901 French made a concerted effort to drive the still-active remnants of the Swartkop commando from their stronghold. Weber described the ominous advance: 'Like cloud-shadows, the dark masses of French's cavalry were passing down the long mountain slopes [from Krugersdorp]. More kept appearing on the roads above. They then spread out like a fan on the slopes. Their scouts had already disappeared in the trees of Sterkfontein' (Weber n.d.: 23).

Fig. 14.8 Boer graves and monument on 'Tweefontein'

Day after day the British shelled the farms in the area from the Krugersdorp ridge. Just as tenaciously, the Boers defended their property as cavalry and artillery swept the hills and valleys from Krugersdorp to Broederstroom. The British did not have an easy task. Skirmishes were common – there was one in the Cradle area at the entrance to the mine near 'Kromdraai'. On another occasion, Major RA Browne and Lance-Corporal Turnbull were mentioned in dispatches for their courage in entering a cave in order to capture six armed Boers. By the end of April the stronghold had been taken, but even after the commandos had finally been driven out, this district required constant policing and vigilance.

The tortuous end

In the final years of the war neither side could find a way to end the debilitating struggle. Exhausted and faced with diminishing resources, the Boers abandoned any hope of victory but persisted with hit-and-run tactics in the hope that their enemy would eventually succumb to frustration and lack of political support. The British response was to try to strangle Boer resistance in a countrywide web of barbed-wire enclosures, with blockhouse forts and booby traps spaced every kilometre along the fence. Manning the blockhouse system overstretched army personnel and in London the Parliament of the day was obsessed with whether black Africans were being employed to guard the blockhouses. Modern historians have concluded, not surprisingly, that in some instances they were (Warwick 1983), but Lord Kitchener always evaded the question for fear of disturbing the Victorian notion that this was a white man's war.

The Cradle area fell in the centre of one such enclosure, bounded by the Magaliesberg in the north, the Johannesburg–Pretoria railway in the east, the Johannesburg–Potchefstroom line in the south and a chain of blockhouses across the western highveld from Frederiksdal to Olifantsnek near Rustenburg. Within this vast corral, the South African Constabulary (SAC), a special force under Colonel Baden-Powell of Mafikeng fame, maintained British law and military presence. SAC outposts were scattered over the Bankenveld, especially along the main routes and mountain passes. They functioned as deterrents rather than defensive positions, and few saw any military action. Most are now tumbled ruins but one, known as 'Barton's Folly', is well preserved and prominent at the foot of the pass near Hekpoort. It stands as one of the few structural reminders of the bitter battles and bloodshed that shaped South African destiny in what is now the 'Cradle of Humankind'.

CHAPTER 15

WHITE SOUTH AFRICA'S 'WEAK SONS': POOR WHITES AND THE HARTBEESPOORT DAM

Tim Clynick

The Hartbeespoort Dam today presents itself as a middle-class (white) playground. The description posted on a number of internet sites runs more or less as follows:

> It is popular for a range of water sports and has a number of clubs – the most active ones are the Marathon Club which hosts an annual '*Om die Dam*' marathon in August – and a number of golfing estates [Pecanwood Estate for example]. It also has a music farm labelled 'Saloon Route 66' which has been dubbed the home of country music in Africa. It has a number of property developments and timeshare facilities. It caters predominantly for the middle classes and is a popular area for suburbanites from Johannesburg and Pretoria residents.[1]

Not much else about the dam and its environs readily comes to hand for the average visitor, whether foreign tourist or South African. All public information about its origins has been effaced and any sense of its true significance for early and mid-twentieth century South Africa has been comprehensively lost. Yet Hartbeespoort Dam in reality occupies a pivotal and highly symbolic role in the history of white South Africa in particular, standing as it did (but obviously no longer does) as a mute monument to systematised racial discrimination and domination. The primary purpose of the dam in its original conception was to contribute towards a solution of South Africa's ever increasing 'poor-white' problem'. Like so much else in this period of South African history, the idea behind the dam's construction is intimately connected with former Prime Minister JC Smuts. Subsequently, key politicians and government officials put Smuts's idea into practice. This chapter explores this story of the Hartbeespoort Dam as the crucible for state policies towards South Africa's poor whites during the first half of the twentieth century.

The weakest link in the chain

The growth of a class of poor whites first emerged as a serious source of concern in South Africa in the 1890s, especially once the arrival of railways in the interior around the middle of that decade, and the 1896 *rinderpest* epidemic which killed up to 90 per cent of the cattle population of South Africa, destroyed the lucrative occupation of transport riding hitherto resorted to by many landless and other whites. The South African War of 1899–1902, and especially the devastation wrought in the Orange Free State and the Transvaal by the British army generals Roberts and Kitchener, massively expanded the number of poor whites. In 1903. 10 000 white indigents remained marooned in the concentration camps, unable to return to *bywoner* (tenant farmer) life. At that point Alfred Lord Milner, the British high commissioner in South Africa, expressed a view which would be articulated again and again in South African political circles over the next forty years. 'The position of poor whites among the vastly more numerous black population,' he observed, 'requires that even their lowest ranks should be able to maintain a standard of living above that of the poorest sections of the population of a purely white country' (Transvaal Indigency Commission 1908: 58). Alternating droughts and economic depressions over the next five years led to the establishment in the Transvaal Colony of the Transvaal Indigency Commission, which presented its report in 1908. The report made a direct link between the declining situation of a significant portion of the white population and the small but marked increase in the prosperity of a section of South Africa's blacks. 'If the white race is to remain the dominant power in South Africa,' this document opined, 'it must live to the standard set by the civilisation it represents. How important it is that the indigent classes should become independent and self supporting may be seen from the fact that while they are falling, the native is actually rising in the social scale' (Transvaal Indigency Commission 1908: 47).

Fig. 15.1 Wagons transporting building materials to the dam site

The Landless Man on the Manless Land
The Great Closer Settlement Scheme at Brits
By JAMES G. McQUADE.

If you want to see where the desert blossoms like the rose, go to Brits. Seven years ago the land below the great dam at Hartebeestpoort, in the spacious valley of the Crocodile River, was virgin soil which had never been touched by the plough, and covered by stunted bush. To-day it is the home of 614 farmers and their families who are producing crops of wheat, tobacco, oathay and lucerne of an estimated value of £80,000. This result has been achieved through the agency of the Department of Lands at a cost of some £1,600,000 or roughly £750 to £1,500 per settler.

South Africa, it has been stated with justice, has the most generous land laws in the world. The amending act of 1924, which was the coping stone on the legislation of 1912, certainly seemed the last word in liberality. But it was capped by the Act of 1931, which placed settlement on a basis of 40 years purchase.

The genesis of the Hartebeestpoort settlement was the probationary training scheme started in 1924 in the Crocodile Valley by the present Minister of Lands, Mr. Piet Grobler. Here the prospective farmer may start with less than nothing. That is to say he can get ground on leasehold, be instructed in the arts of husbandry, have a house built on his holding, obtain implements and stock on loan, and be paid a subsistence allowance while he is establishing himself.

Originally the Hartebeestpoort Probationary settlement scheme was intended to arrest the drift to the towns of the rural population. It has become one of the most successful closer settlement schemes in existence. It aims at bringing the landless man to the manless land; at reviving the agricultural instinct of the rural labourer: at recapturing his "land" sense, his spirit of industry and independence before it is completely crushed in the remorseless mill of modern town squalor. The material on which it works is the descendant of the dispossessed landowner, who through a succession of misfortunes due to war, drought and pestilence, is in danger of losing his industrial sense and sinking below the bread line.

The scene of this inspiring experiment has been admirably chosen. A two hours exhilarating spin by car from Johannesburg takes one to the administrative offices of the Controller, at Brits, now a thriving township which is the pivot of the settlement. The writer in company with a fellow journalist recently had the opportunity of going over the ground and enjoying the gracious hospitality which Mr. P. B. van Rhyn dispenses with traditional Boer courtesy in the cosy homestead nestling in the hollow of a verdure-clad kopje overlooking the smiling valley which, on the day of our visit lay bathed in the golden sunlight of a bright summer morning.

The Minister is fortunate in his Controller. Mr. Van Rhyn is a farmer of the enterprising Voortrekker type. He regards the settlement as one big farm, and the settlers as in a sense his family. On them he imposes a Spartan-like discipline tempered with paternal indulgence towards human frailty. But he exacts implicit obedience to the rules of the household, in this case the statutory regulations. These by no means err on the side of slackness or leniency. The sun is his clock, and more often than not he beats the clock in the race to greet the early bird. Woe to the worm which ignores his embarrassing punctuality. From the broad stoep of his house he can see the blue smoke curling up from every cottage and a prompt and sharp reprimand is ready for the sluggard who still lies abed while his fellows are in the fields.

Up to date 452 settlers have been trained and placed on the land under the original probationary scheme, and may be regarded as re-established farmers who have made good. Most of them are making a living, and some are now in comfortable circumstances and are able to make their redemption payments freely and regularly. There are another 162 in training. So successful does the Department regard the scheme that an additional £10,000 has been voted for the purchase of milch cows and lucerne seed. To date 164 dairy cows have been bought and lucerne planted. This will lead to the introduction of dairy farming as an additional line, the Department having constructed seven new cattle dips for the use of the settlers, a further 162

L

357

Fig. 15.2 A contemporary article praising 'one of the most successful
closer settlement schemes in existence'

A new cycle of depression and unemployment prior to World War One prompted the formation of the Unemployment Commission of 1913, and in due course led to Parliament voting funds for the Hartbeespoort Dam project in 1914. The Hartbeespoort Dam – situated on the Crocodile River in the midst of the Magaliesberg mountain range, at the doorstep of the 'Cradle of Humankind' – was by far the most ambitious irrigation project to be undertaken by the South African government in the 1910s and 1920s. Its prime purpose was to mop up and rehabilitate 'poor whites'. Work on the project was initially delayed by the First World War, but the 1916 drought, the severest in decades, which pushed thousands of whites off the land and into the Witwatersrand's inner-city, multiracial slums, led to the reactivation of plans to construct the dam, on which real work began in 1919. Another massive countrywide drought in 1919, which had similar effects to that of its predecessor in 1916, resulted in the government deciding to adopt an all-white employment policy on its irrigation projects, particularly the Hartbeespoort Dam. From this point on the Hartbeespoort Dam became the government's largest and most ambitious poor-white project by far. By August 1922 it employed a massive 3 400 white workers, and supported a total community, including their dependents, of 10 000.

In 1919 the Irrigation Department approved a new scheme for a permanent poor-white farming community whose land would be irrigated by canals below the dam. Concern mounted again in 1924, after the renewed onset of depression and the most severe drought and locust invasion for years, when the Smuts government was somewhat unfairly accused of not seriously confronting the poor-white problem. In that year Smuts called a general election. In what shortly became dubbed 'The Poor White Election', he lost to a Nationalist/Labour Pact group led by General JB Hertzog and Colonel FHP Creswell. The basic strategy of the new government for dealing with the problem that emerged after 1924 was to initiate a 'training and sorting' process among ex-Hartbeespoort Dam construction workers, whereby the more disciplined and successful of them would be allowed over time to earn the right to freehold at Hartbeespoort after successfully working the government-owned smallholdings on Losperfontein, while failures would be sent to compulsory farm colonies.

The Pact government: From poor-white relief to rehabilitation

The new government was composed of Afrikaner nationalists and English workers. According to a contemporary, the Pact mandate was to resolve once and for all whether South Africa was to become 'a huge black compound for the big capitalists or a prosperous white man's country'(Hancock 1968: 162). The Pact was supported *en masse* by white workers who in 1922 had engaged in violent protest and insurrection against Smuts and the mine owners. The creation of the so-called 'colour bar', to protect the jobs of whites in urban and industrial professions, was a direct result of their activities.

The new government also introduced a radical new proposal in respect of rural poor whites – this was for a state-driven programme to rehabilitate white South Africa's 'weak sons' and to find for them a permanent place in the rural economy. Responsibility for this programme was put into the hands of the Department of Labour, with Colonel FHP Creswell, the leader of the South African Labour Party, as its new minister. Minister Creswell stated that the solution was 'simple and clear – to re-organise the rural economy so as to find a place in it for white people' (ACL 23–24 September 1924: Statement by Minister Creswell). The department's first secretary, CW Cousins, announced that the department was established so that '[c]ivilised man in South Africa will assume conscious control and work out his destiny' (ACL 23–24 September 1924: Statement by CW Cousins).

Previous governments had sponsored relief work for the white poor on public works, together with pauper doles to support the large 'floating population' of unem-

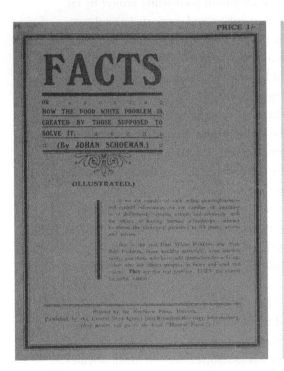

Fig. 15.3 *The cover and first page of* Facts, *Johan Schoeman's pamphlet condemning the Losperfontein Probationary Settlement*

ployed and unemployable whites. One such monument to these early public relief efforts already stood in the Cradle area – the Hartbeespoort Dam and associated irrigation settlement. But Minister Creswell worried that rural relief work of this kind was, in his words, 'demoralising, unconstructive and unsatisfactory'. Many poor-white relief workers objected that the work was only of temporary duration – but didn't the government owe them a permanent living? Others complained that they were only given work on rural public works so as to keep them out of the cities. A government advisory body of the time claimed in turn that 'relief workers in the towns would not go back to the land on their own initiative; they required to be sent back'.

The Pact government took the view that previous governments had had a hand in creating this poor-white underclass by failing to stand up to it. A new 'white labour policy' was required that would utilise the great public works programmes – such as the massive irrigation settlement that was built by the Pact downstream of the Hartbeespoort Dam wall – to rehabilitate once and for all poor whites from the rural areas, turning them into self sufficient citizens in a 'white man's country'. The Pact would therefore formulate a 'civilised labour policy' for the South African countryside.

The Pact's programme of reform was quite novel in South African history. It was clearly grounded in the new imperial 'science' of eugenics. Poor whites were seen as having lost 'racial pride'. This pride could only be restored by providing the poor and marginalised with 'honest opportunities' for work – or as one reformer put it, to 'replace the "locust" principle of the poor white with the "ant" principle of the industrious working man'. Rural Afrikaners would be 'put back on their feet'. This would allow the urban white workers to protect their labour markets from being undercut. Afrikaner nationalists and English trade union leaders therefore found common purpose in respect of the policies required to rehabilitate poor whites.

The Pact's programme to 'rescue white civilisation' in South Africa implied a new and highly coercive and intrusive role for state officials in rehabilitating white people who were poor. State intervention massively expanded in this inter-war period. Historians have noted how this changing role of the state impacted on black South Africans; what is less often realised is how the state and its officials turned their attentions to poor whites during this crucial period. For example, the Pact was the first government to pass legislation allowing state officials to commit individuals who refused to accept offers of work ('won't works') to labour colonies.

This new departure in South African public policy took place against the backdrop of *two* contemporary political problems: the so-called 'native question' *and* the 'poor-white question'. Both poor whites and blacks from the rural areas lacked the necessary

*Fig. 15.4 A panoramic view of the new Hartbeespoort Dam recreational site,
published in a promotional brochure in 1926*

skills to prosper in South Africa's cities and towns. Both were predominantly from marginal rural communities located on the periphery of the urban industrial economy, with few property rights and living on land owned by others. White politicians realised that, in order to prevent poor whites and blacks from merging into a great (and dangerous) amorphous underclass, they needed a programme that would separate the fortunes of the white and black rural poor in the Union. The Pact government promoted the 'white labour principle' that would give opportunities to poor whites, whilst pushing black South Africans out of the industrial economy except as providers of migrant labour, and pegging them back into rural reserve areas. For this programme to succeed, however, the white poor needed to be taught the 'dignity of work' as well as to 'stand on their own two feet'. They could not be allowed to wallow in the pre-industrial past, where force and coercion allowed many whites to live off the land and labour of Africans. The government had to play the role of 'father' and 'mentor' to poor whites, and it was at the Hartbeespoort Dam that they worked out how they would play those roles.

Losperfontein (Hartbeespoort Dam) Probationary Settlement for poor whites

The new programme for poor whites was first put into practice on the Department of Labour's new Agricultural Training Farm, which was located on a portion of the

Hartbeespoort Dam Irrigation Scheme called Losperfontein. As already noted, the dam and a large irrigation settlement were constructed by 1924. With the accession of the Pact – and the creation of the new governmental Department of Labour – it was decided to expand the number of probationary settlers who would be drawn from the ranks of poor whites. Some 14 700 morgen of land within the settlement on the left bank of the Crocodile River were handed over to the Department of Labour for this purpose. The area was called Losperfontein, and all probationary settlers were put under the direct control of officials of the Department of Labour. In July 1924 the Labour Department assumed control over the probationary settlers as well as over the remaining white dam and irrigation settlement workers and their families (1 657 workers and their families – a total of 4 777 people). Potential settlers at Hartbeespoort who had their own resources made application directly to the Irrigation Department, which retained overall control of the Hartbeespoort Dam and the remaining land on the settlement. Land was now theoretically available to three different kinds of settlers:

- settlers with small capital under the ordinary conditions of the Land Settlement Act, 1912;
- settlers who could put down a portion of the purchase price in cash under the pro-visions of the same Act as amended in particular by Section 7 of Act 21 of 1922;
- probationary settlers without regard to capital or experience who would be required to undergo a period of training before being allotted holdings under the Land Settlement Act.

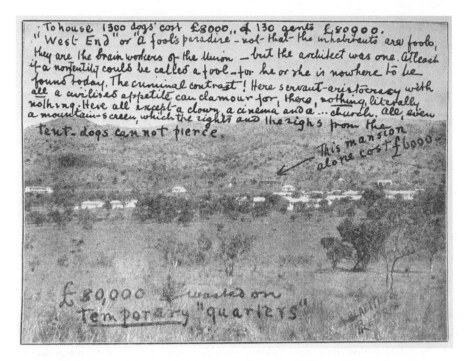

Fig. 15.5 Annotated photograph of the accommodation provided for the 'brainworkers of the Union', from Johan Schoeman's pamphlet Facts

The Department of Labour was now in a position to proceed with its 'white labour' experiment at Losperfontein. Cousins described their objectives thus:

> The main object of the department is…while making the most advantageous use of the material possibilities of the farm as an economic proposition, to attempt to use the scheme…for the rehabilitation and reinstatement on the land of the class of men generally known as the 'poor white'. The men actually dealt with on the farm will be men whom no farmer is willing to employ in their present state, who have flocked onto Government relief works whenever these were available, and who have frequently been described as incapable of taking a reasonable share in the production of the country (ARB. Box 205, LB510/26. March 1925. Hartebeestpoort Area. Departmental Committee. Memorandum).

Cousins described how this objective of rehabilitating poor whites and reinstating them 'on the land' would involve not just the breadwinner but the entire poor-white family. He wrote:

It is the object of the Department to turn out not only men who have been trained, but also to send out from the settlement whole families as social units who will make better material for land settlement than is possible under any scheme hitherto practised or proposed. This can only be done if the Department is placed in a position to have complete disciplinary control of the people employed, so as to direct their lives into useful channels. It is impossible to dissociate the control of the work and the control of the men who perform it. This is the essence of the problem for which a solution is sought.

By 1927, so enamoured was Cousins of the progress made at Losperfontein that he was 'convinced…that they had found the key to the situation, but why stop there?' Under his guidance the department then began incorporating key elements of the probationary programme developed at Losperfontein into nearly every public works programme in the country. Losperfontein Training Farm became the flagship project of the Pact government for dealing with poor whites.

What were the key elements of what Cousins described as 'the machinery' developed at Losperfontein? Firstly, the 'native factor' was outlawed altogether (*Government Gazette* 1924; note that all further references to 'the regulations' are from this source.). 'All native servants have been dispensed with and the labourers have to depend entirely upon themselves without native assistance,' read the regulations. This principle of white labour only had been in place on the Hartbeespoort Dam project from its inception. White relief workers constructed the dam, carved out the irrigation canals, cleared and levelled the land, and laid out the road network. Nevertheless, probationers and department officials clashed continually over this principle. The Losperfontein superintendent, for example, complained in 1929 that whilst Losperfontein was supposed to be a white settlement, this was not perfectly true:

He would like it to be so but certain inhabitants wished to use native labour for washing and charring. He had explained that these people should do the work themselves or, in the event of illness, should get their children to do it or enter into an arrangement with their neighbours. In connection with the workers it was possible to insist on no native labour being employed but difficulty was experienced with the school teachers…The administrative staff [was] prepared to use white labour if it could be obtained but the European girls on the settlement refused [to work]. He had tried to get a certain girl to undertake such work and she had agreed provided she did not have to clean the stove or cut wood and provided she received £5 a month. The men could not afford to pay this wage with the result that they had to use native labour (ACL 20–21 April 1927: Statement by CW Cousins).

Labour Department officials lamented these 'lapses' on the part of probationary settlers to undertake all manual work on the settlement. This was exactly why a new policy regarding the poor whites was required. The views of an inspector of the Labour Department, E Creswell, vividly capture this zeal and sense of mission amongst government officials on the settlement. In a report to the Advisory Council of Labour on 'Farm and Labour Colonies' he wrote:

> In every case where…the views of responsible men in the Rural areas on the subject of the Poor Whites [was requested] the answer always worked out the same. 'They are hopeless; nothing can be done with them'…[But if] these people were given ground, seed, implements and oxen, and placed under the supervision of some official who would tell him how to set about working the ground and leave him, and return back after a period and see how things were going since his last visit, it would meet the case (ACL 20–21 April 1927: Statement by Inspector Creswell on 'Farm and Labour Colonies'. All further quotes attributed to Inspector Creswell in the following section are from this source.).

Inspector Creswell believed that the South African case was 'infinitely more complex than the position of the unemployed in Great Britain. In South Africa there were various standards and [various] races [to be taken into account] when seeking for means to preserve the White Races from decadency'.

It was these unique colonial circumstances that justified drawing a rigid, unbreakable barrier between 'civilised and uncivilised standards of life':

> The fact must always be borne in mind that our efforts are directed to reclaim these people; in [my] scheme we are putting them to a test, for there is no alternative if they fail to seize the opportunity but the Labour Colony. They cannot fall back upon the casual relief and charity which prevails today. In trying to assist them to reconstruct their lives and provide them with something more than a blank wall and raise them above conditions seething with degradation, we must not place any temptation on [sic] their way once they have secured a foothold in allowing a bad season or two to force them to seek aid from some financial source to carry on and perhaps not too good a season following, find them unable to meet their financial obligations, see the Bondholder step in and take all [sic]. You will see that I realise very drastic measures will have to be adopted if these people are to be saved.

Inspector Creswell believed that poor whites should not be permitted to become owners of land in their lifetime:

When drought came they would get into distress and fall into the hands of the speculator, who would take control of the situation; and this would eventually place the poor whites back in the position in which they were before. [They]… should not be placed on the land direct: there should be a co-operative system, which would keep the outsider out. The people concerned should be a little community on their own. If a man desired the ownership of a farm, he had the Land Settlement Act; and, if he proved himself to have the necessary qualities, he could make good in that direction. The place which he envisaged for poor whites in the collective sense was a sort of 'training ground'.

For whites to own land, the right to property ownership must be earned. On state-run farm colonies, officials could ensure that poor-white families were taught the responsibilities that came with ownership and agricultural enterprise. They needed to be instructed in 'modern methods' of intensive farming. This official concluded that these steps were necessary in order to 'to cut away that dependence upon the Native for many forms of labour looked upon today as the province of the native…I want to see those given a chance willing to work and in the process of the scheme weed out the unemployables.'

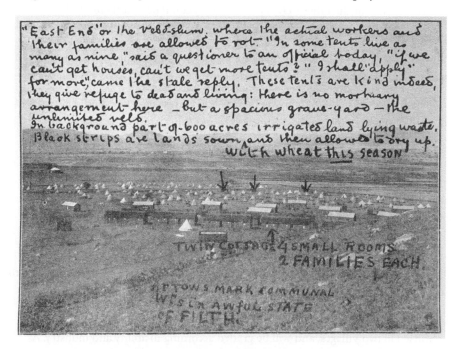

Fig. 15.6 Annotated photograph of accommodation in the 'Veld-slum where the "actual workers" lived', from Johan Schoeman's pamphlet Facts

Behind the motives of rescue and rehabilitation lay the threat of state coercion and control of the lives of individuals by officials. It was Minister Creswell who presided over legislation that made penal sanction for 'won't works' possible for the first time: the Works Colony Act No. XX of 1927. This was necessary, he stated, since '[amongst] poor whites there would always be found a number, although physically and mentally capable of work always malingering and preying upon the genuine worker' (ACL 14–15 December 1925: Statement by Reverend van der Horst). Officials required the powers necessary to detain any such malingerers, even indefinitely, until such time as they had mended their ways. This meant that '[t]he liberty of such persons shall be restricted and no sloppy sentiment shall be permitted to interfere with the Administration'. The state now possessed the right to punish recidivist 'won't works' in pursuit of its mission to save poor whites. This was a boon for officials on Losperfontein. The superintendent of the settlement, Reverend van der Horst, put it this way:

> Without the threat of commitment to a work colony all he could do to disci-
> pline backsliders was to reduce the man's wages and to try and make him feel
> that he had done wrong…[Therefore he] wanted a Work Colony so that if the
> men did not conduct themselves properly they could be sent there (ACL 14–15
> December 1925: Statement by Reverend van der Horst).

Welfare officer Reverend Fick felt that the prospect of commitment to a work colony would really be 'a tremendous help to these people':

> [They] were too good for a prison but too bad for society. They should be treat-
> ed like naughty children and forced to work; [but] in the end they would be
> grateful…They could never take away the control over those people. If they did,
> they would go under again (ACL 14–15 December 1925: Statement by
> Reverend Fick).

Rehabilitating poor whites

Poor white probationers were put through a series of tests – hurdles – on their way to complete rehabilitation, which in essence meant that the Secretary of Labour would event-ually grant them the right to purchase a plot on the Hartbeespoort irrigation scheme itself. Initially the programme at Hartbeespoort Dam could be described very simply:

> These men will be under discipline, will be paid certain wage rates, will be
> housed in two villages, and will be graded according to ability and industry. As
> men prove themselves qualified they will be passed out from the training farm
> with a view to becoming tenant farmers under the Lands Department. Men

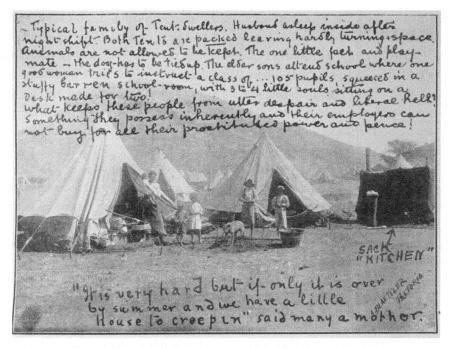

Fig. 15.7 Annotated photograph of living conditions for 'a typical family of tent-dwellers', from Johan Schoeman's pamphlet Facts

who are incapable of benefiting by the training will be passed out from the training farm to a farm or labour colony of another character, or into other avenues (ARB. March 1925. Hartebeestpoort Area. Departmental Committee. Memorandum).

However, a more elaborate 'scientific' programme consisting of four tiers soon emerged, guided by inputs from the superintendent and his welfare officers.

The first tier provided for a selection process from the undifferentiated mass of individual poor whites that demonstrated their 'readiness' to be reformed. Officials 'sorted out' the 'won't works' from the 'will works'. In the words of a Losperfontein official:

> They afford a means of weeding out persons of certain classes – the unemployable, phthisis men, the human wreckage of the country, persons who had no pensions, persons who were at present a burden on charitable institutions, and so on. Men who had been on the land should, if possible, be put back on the land; and people who would not work should be diverted to labour colonies (ACL 23–24 September 1924: Statement by E Creswell).

Only those poor whites who successfully saw through a period of relief labour were adjudged to have embraced rehabilitation and the dignity of work. Many failed the test: they deserted from the relief works in great numbers since the work was heavy and took place under less than ideal conditions. The forestry settlements were considered the worst open-air relief option, leaving men physically broken after as little as six months' labour. Those who passed this demanding test were presented with application forms by department officials to apply for a place on the probationary settlement at Losperfontein.

Records were centralised, cross-referenced and sorted to prevent duplication or abuse by potential applicants. All recipients of relief labour in the country – including those on the national road work development programme, the forestry settlements, municipal or provincial construction work or the like – were carefully graded and evaluated by officials in terms of their health, age, social characteristics, marital status, numbers of dependents, reliability, willingness to work, enthusiasm and initiative. Applicants from towns and cities were bypassed in favour of rural poor whites. Secretary Cousins reported in 1925 that:

> [t]he labour camps at Hartebeestpoort enable the ARB to sort out and classify the various men engaged. Some, undoubtedly, will make good ultimately as small-holders under any land-settlement scheme; others, though not possess-ing the character to fit them to be holders of land, may nevertheless be found very useful as labourers or foremen on irrigation settlements, while others will not be of use in farming and will have to be relegated to navvy work pure and simple, or be dealt with otherwise (ACL 14–15 December 1925: Statement by Reverend van der Horst).

As the rural depression of the latter 1920s deepened, the Department of Labour was inundated with applications for the Hartbeespoort probationary settlement by relief workers from all parts of the country, so desperate were many to gain access to state support. Initially, however, the department drew its probationary settlers from relief workers from the Hartbeespoort/'Cradle of Humankind' area.

The second tier of the programme placed probationers in an intermediate position between independent (unsupervised) and supervised work (what was called 'commu-nal' labour), both performed at a minimal daily wage. Probationers were provided with family housing, fuel, water, and medical services for which they were required to con-tribute a portion of the costs. Probationers then progressed from work in gangs on the farm, under the direct supervision of the farm manager, to independent work on a three-morgen garden or on vegetable plots. This phase lasted approximately eighteen months. Probationers were paid a daily wage of 5/6 for this period. As their work

Fig. 15.8 Annotated photograph of the polluted water supply at Losperfontein,
from Johan Schoeman's pamphlet Facts

progressed to independent work on their small plots, profits made from the sale of produce were credited to an account kept in trust for them by officials.

This second, intermediate level was deemed crucial by reformers for weeding out
those who were not serious about being rehabilitated. There were still many poor whites
who arrived at Losperfontein with what the superintendent referred to as 'the spirit of
movement' (ACL 14–15 December 1925: Statement by Reverend van der Horst). This
second tier was designed to sort out these men from 'the better class [who] are anxious
to acquire land. Once this was completed officials could teach this better class [of poor
whites to] work the ground on their own, and demonstrate to them that by working a
small plot of ground they could earn far more than the daily [relief workers'] wage of
5/- to 6/- per day. If they failed [to learn this lesson] they would be returned to farm
labour' (ACL 12 November 1928: Statement by Reverend van der Horst).

An unusual number of poor whites undergoing the gruelling programme established on the settlement rejected the kinds of lessons that officials felt they ought to be
learning. 'These people had lost their sense of responsibility,' explained the supervisor:

> They were willing to work but under the wing of the Government. They wanted
> to work for wages, and they feared the uncertainty of farming. There was great

difficulty in getting men to move away from casual or road work for the purpose of taking up farm work. It was even difficult to induce them into the tenant farmer extension schemes. They preferred to get a wage of which they were certain (ACL 12 November 1928: Statement by Reverend van der Horst).

Levels three and four of the programme meshed contemporary advances in agricultural science with a new 'science' of white labour in a truly innovative way. A new productive regime was instituted that comprehensively altered the nature of land settlement in the country. This was the idea of intensive farming on small plots, a new agricultural small-holder regime. Probationers were instructed in 'sound farming practices' that departed fundamentally from traditional farming methods such as labour tenancy or more commonly 'farming on the half' for landowners, for example. These wasteful methods of farming in the past were contrasted with 'modern' intensive (irrigated) cash-crop farming on small plots. In addition, probationers were compelled to work without labour inputs from black workers, and to abandon traditional means of obtaining draught animals, seed and equipment (such as by hocking future harvests to rural traders – who were often Indian and Jewish – against book loans for agricultural supplies at high rates of interest).

The main features of this system, the Department of Agriculture's advisors explained, were that 'the settler and his family should work his own holding without calling in native labour, without falling back on the State for running capital, and by [co-operating amongst themselves] on how to share seeds and breeding stock' (ARB Box 205, 14 August 1925: 'Experimental Station and Training of Probationary Settlers'. All the information quoted on agriculture in the following section is from this source.). Intensive smallholdings of this kind enabled white farmers to:

- spread risks, by having many revenue producing lines;
- preserve soil fertility by introducing 'the animal factor' onto the smallholding;
- utilise both animal and man on the plot throughout the year and avoid 'rushes of work at any particular season of the year that in the past necessitated the employment of outside (black) labour and, possibly, the hiring of animals';
- cooperate with fellow settlers so as to breed a spirit of self-reliance, of independence and self-sufficiency.

An important aspect of this strategy was to establish an optimum size of 'economic unit' that a settler and his family could successfully work without falling back into 'old habits'. In this way harmony between physical occupation of space, use of farming equipment and other productive inputs, and the basic requirements for self-sufficiency amongst self-supporting white nuclear families would be achieved.

Official 'planning' would therefore begin to control the natural environment, for

Fig. 15.9 Annotated photograph of the water source for 'people and natives' at Losperfontein, from Johan Schoeman's pamphlet Facts

were not poor whites the products of a way of life too close to nature, and guilty of not sufficiently applying the lessons of modern science to their environment? Officials were constantly scanning and measuring the relationships between space, enterprise and profitability on the settlement. 'If the holding is too small,' ran one account, 'the capital expenditure in the shape of houses, outbuildings, fencing, domestic water supply, etc. is spread over the holding, [and] it means that the area is over capitalised before the settler starts.' There had to be economy in how capital inputs related to output (production). Likewise, the families of probationers were input factors that were critical in creating self-sufficient, closed, productive units. This vision of the modern white colonial farm family contrasted markedly with what officials saw in poor-white families, with their phalanxes of distantly related and multigenerational family members. The ideal farm family was a nuclear one; probationers needed to be prohibited from supporting any extended kin not directly part of the nuclear unit. Within nuclear families each family member needed to be viewed as a unique productive unit: no one member could live 'off' the labour of another. 'The men, women and children were all working – that was the outstanding success of the scheme,' explained an official.

This integrated menu of inputs was required to scientifically sustain an intensively worked, highly productive piece of irrigable land. Agricultural officer RW Thornton stated:

> If the holding is too large, it means that the settler and his family, and working in co-operation with his neighbours, cannot work [it] economically. This leads to bad farming and bad farming practice.

Finely calibrated inputs of labour power, capital and collective effort met in glorious symbiosis in the ideal of a colonial 'smallholder'. Every item on the probationers' plot was carefully itemised and a financial value placed on it by officials, who maintained strict individual cost accounts. The 'profit and loss profile' of each smallholding, of each trainee family, was constantly weighed and then subjected to official examination. The elimination of any 'wastage' in terms of inputs was critical to the achievement of productive efficiency. Thus the poor white spirit of waste and squander would be countered by the 'smallholder' programme. Through this programme, noted Colonel Muller, department officials 'instil the idea that farming is an economic proposition…They were [therefore] out to get everything on sound business lines, and the right type of man…selected. Losperfontein [was not] a dumping ground for any sort of man' (ACL 17 April 1926: Statement by Colonel Muller).

The kinds of crops planted by probationers were restricted by officials to those which best supported the overall goals of efficiency and productivity. Simple rotational practices with new crops with high cash returns, like tobacco and winter wheat (sugar beet was briefly mentioned as a possible alternative), were encouraged in conjunction with small-scale dairy farming. Important sidelines under the management of wives and daughters, such as poultry and pig breeding, vegetable plots and fruit orchards and the like, were encouraged to achieve productive on-farm self-sufficiency for all family members. These new agricultural practices would also be a counterweight to the 'spirit of movement' amongst poor whites in this period of South Africa's history which officials felt contributed to profligacy and waste – for example, seasonal off-farm migration by men and boys to the diamond diggings and by women and girls to the citrus packing houses and tobacco factories, where cash wages could be earned in relatively short periods of time.

Coercion and control of poor whites

The new labour regime demanded of probationers by the smallholder strategy was enforced by a thicket of regulations, and the often arbitrary exercise of authority by officials at Losperfontein. '[The] settlers are under strict supervision and have to comply with certain stringent regulations which govern the Settlement,' began Government

Notice 1957 of 1924. Clear lines of command ran from the local welfare officer, who inspected and implemented the regulations, to the superintendent, who was responsible for all decisions taken. 'We are slaves under the eye of the *Dominee*,' wrote JJ Britz. 'If you won't take up the *graaf* and *pik* [shovel and pick], then you must hit the road. Why must I be treated like a white *Kaffir*…I am an old burger of the country and have been shot to bits in its service. Why should I be treated like this?' (ARB. Box 209, LB 510/11, 7 June 1926. Part 3. Complaints. Relief Works. Hartebeestpoort. Letter JJ Brits).

Welfare officers were responsible for the direct supervision, instruction and general welfare of tenants and their families. The regulations stated that they had the right at all times 'to inspect the dwellings, gardens etc., of tenants in order to see that the regulations and instructions have been complied with'. With this responsibility came the power to suspend any probationer for 'drunkenness, gambling, disorderly conduct, immorality, insubordination, or other misconduct' subject to the review of the superintendent, who would then 'determine whether a tenant is to be dismissed or what action, if any, should be taken against him'.

Probationers' rights extended only so far as 'the right to occupy temporarily a plot of land on a farm selected and acquired by the Department of Labour'. Probationers' duties were extremely onerous: together with their families they were 'bound to carry out any instructions which may be given by [the welfare officer] in the interests of the tenants and the farm'. Males heading up families – described as 'tenants' – were responsible for the 'good conduct of the members of their household'. Together with their wives, they signed a memorandum of agreement to submit themselves and their children to the supervision and regulation of a welfare officer. In doing so they agreed to allow the welfare officer to compile an inventory of all financial and material advances made to the family by the department. The agreement also extended to seeking the '[p]rior permission of the Supervising Welfare Officer' before using any money to purchase any article for the household. Each tenant was also held responsible for 'the effective cultivation of the holding leased to him and for the proper care of buildings, implements, livestock etc. …provided for his use'.

A system of cash advances – related to the size of the tenant's family – bound the tenant to the authority of the welfare officer. In the event of non-compliance with the regulations and conditions of employment, 'The Secretary [of the Department of Labour], acting on the recommendations of the Welfare Officer, may reduce the amount of the advance to any tenant who contravenes any regulation or who fails to carry out any lawful instructions or who is guilty of any misconduct.' Thus entire families were punished for any transgressions by the probationer. Regulation 11 designated contravention of the memorandum of agreement – disobeying the instructions of a welfare officer or failing to stay permanently on the plot – an administrative crime, allowing the welfare officer the right to confiscate 'either wholly or in part' the tenant's interests in any agricultural and pastoral produce or any monies held in trust for the tenant or his family. A behavioural code was also embedded in the regulations: 'No tenant or member of his family may employ Natives or Coloured persons,' ran Regulation 14. Regulation 15 stated that no tenant 'may make, sell, keep or use on the farm intoxicating liquor'. Regulation 17 prohibited tenants or members of their family from '[absenting] himself or herself from the farm without written permission from the Welfare Officer'. Tenants were in addition not permitted to accommodate visitors or to remove departmental equipment or livestock from the farm without consent of the welfare officer.

The regulations controlling the marketing of the crops of probationary trainees also reflected the patronising and controlling content of the rehabilitative programme. Regulation 21 comprehensively outlined the content and rationale of state-promoted co-operative principles and individual accumulation:

> All crops and other products will be marketed by the [Department] on behalf of each group of tenants co-operatively, but each individual producer will be credited with the amount realised by the sale of his or her crops, livestock etc… [All] balances will be held in trust by the Department of Labour until the tenant has accumulated a sufficient sum to enable him to commence farming operations on his own account. No tenant will during the period of occupation be passed out from the farm until the Welfare Officer is satisfied that such tenant is proficient and capable of farming independently. A statement showing the financial position of each tenant will be furnished to him quarterly by the accountant, Department of Labour.

Before accounts were reconciled officials deducted any cash advances made to tenants. They deducted also the probationer's pro-rata share of the total farm rented out to all probationers, together with a depreciation cost of 20 per cent per annum on the value of any departmental implements provided for use by probationers. Water rates, compulsory dipping fees and any other deductions, including those made for disciplinary infractions by probationers or their families, were also provided for in the regulations. Finally, any losses of stock, implements or other goods used communally on the farm were to be a cost borne by the tenants collectively.

It was estimated that it took between three and four years for a probationer to accumulate sufficient cash funds in his personal account for the Secretary for Labour to agree to their release so that the probationer could make a down payment on a government plot of land and take up farming on his own account.

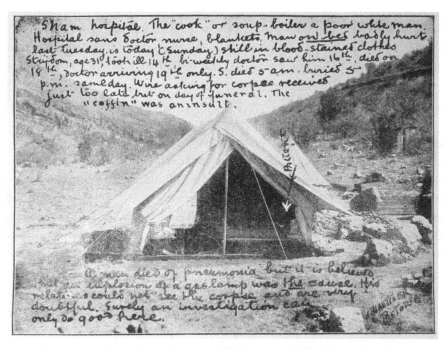

Fig. 15.10 Annotated photograph of the 'sham hospital' facilities at Losperfontein,
from Johan Schoeman's pamphlet Facts

Poor whites and government rehabilitation

The sources record multiple instances of poor families and many individuals resisting the intrusions of officials. Poor whites asked why, simply because they were poor, should they and their families have to be graded and tested before they could get land? Why should the superintendent and his welfare officers be trustees of their savings, if the aim of rehabilitation was to teach financial and social independence? Why should they be forced to save for land at the expense of their present quality of life? Why should they not employ black labour on their plots, if it freed their children for schooling?

It is not possible to explore these aspects of the rehabilitation project in any detail here, but it is clear that dissatisfaction amongst the trainees at Losperfontein was pervasive. A single example should suffice. In 1927 (illegal) meetings were held on Losperfontein, to the chagrin of the superintendent:

> It was a most deplorable affair. Grievances had been aired, and the psychology of the mob was very much in evidence. Discipline was being undermined... they must not interfere with the authority of the officials, who had to maintain the discipline (ACL 20–21 April: Statement by Reverend van der Horst).

269

Disaffection was initially put down largely to 'political influences', to irresponsible troublemakers who 'looked for any excuse to excite government settlers for personal ends'. But by 1928 officials began to suspect that in addition to external troublemakers, factors internal to the probationer families were also making poor whites receptive to such influences.

Officials began detecting, for example, evidence of 'backsliding' in the language that parents used around the home and on their holding, in the attitude of children to their parents, in the degree to which officers detected the 'happiness' of home life, the functionality of the family, the nature of the marital relationship between probationers and their wives. The superintendent already had the right to punish probationers for 'drunkenness, gambling, disorderly conduct'. Now probationers' domestic lives were subject to official inspection and interference (Central Archives Depot, Department of Irrigation. All information quoted on trainees in the following section is from this source.). 'Regulations' were arbitrarily manipulated by officials in order to address supposed domestic and personal weaknesses amongst probationers. Mr AJ van Zyl was classified as 'unpromising' by van der Horst because 'I believe that he only works his plot because he is compelled to do so. His stay here is just for the 4/- *voorskot* [advance] per day. My honest opinion is that he would far rather work for a wage than think about buying land'. HJ Nel was also unpromising: 'He was very untidy in general and thus careless in his actions and also a little lazy and I therefore have very little hope in him.' CB Swart's problem was more complex: not only was he very untidy both on his plot and within the home, and lacked initiative, but his wife was also a very bad influence. She kept a very untidy house, and was herself unkempt. She was a *terugsitting* (handicap) and the superintendent recommended his removal from Losperfontein.

A report on all the Losperfontein probationers dated 1930 explicitly categorised their progress as either 'excellent' or 'moderate' or 'unpromising', based on performance in terms of the following criteria: 'Home conditions'; 'Behaviour of family'; 'General Conduct'. Some probationers proved '[c]apable of making success when allotted plot'. In other cases 'Home conditions' were 'bad, untidy', 'dirty' and 'slovenly'. Comments on the behaviour of the family traversed the range from 'excellent', to '[has] a drink occasionally' or 'Family's life not happy; daughter's character doubtful' and 'Wife quarrelsome; children not well behaved'. Comments regarding the relations between parents and children abound: 'Parents warned re children's conduct' ran one; another, 'Children neglected. Wife objectionable, children attend school irregularly.'

What reformers appear to have been grappling with here was how best to instil amongst poor whites 'self-discipline', so that the probationer and his family would emerge as independent units capable of subsisting independently of further official intervention and support. The regulations governing the settlement, and increasingly the direct inter-ference of reformers themselves, therefore placed inordinate emphasis on the 'natural' patriarchal hierarchy of a fully functional nuclear family and a normative moral code of behaviour. No widows were allowed to stay on the plots unless they had a son to succeed as head of the family. Women were wives and mothers: their roles were as domestic homemakers, and they were not encouraged to become breadwinners, for example. The role of the 'Dutch' churches in policing a new moral code of conduct for poor whites can also be noted at this time, on Losperfontein and more widely in the country.

There were also important generational considerations to be taken into account. A finely calibrated 'opportunity cost' calculation determined the suitability of mature settlers for participation in government rehabilitation schemes. Men of 'mature age' – in their forties and fifties – were excluded by officials from the probationary programme at Losperfontein. Firstly as parents, and then as producers, these poor whites possessed 'set ideas' derived from a lifetime of indigence that officials would find very hard to break. Officials doubted that success would accrue to their efforts, and there would always be the danger of 'backsliding'. 'With persons of mature age the period of training will not be sufficient to fix their ideas in new channels unless means are provided to pre-vent their slipping back,' wrote one official in response to a query from the secretary (ARB Box 205, LB 510/26. 14 August 1925. Experimen-tal Station and Training of Probationary Settlers. RW Thornton, Director of Field and Animal Husbandry, to Under-Secretary Department of Agriculture).

Fig. 15.11 Annotated photograph of a burnt wheat field from which 'starving' families could not take food, from Johan Schoeman's pamphlet Facts

There was a more overwhelming financial consideration: mature settlers would simply not have sufficient working years ahead of them to be productive after they had successfully proceeded through the programme and then begun acquiring land under the Land Settlement Act. The Under-Secretary for Agriculture explained:

With regard to the older adults, except in the minority of cases, all that can be hoped for is an amelioration of the conditions under which they live; [but] with the young people an entire reconstruction of their lives mentally, morally, and materially must be accomplished, otherwise with the very limited means of their parents, and the restricted outlook they must necessarily have, it is felt that their chances of success in life are particularly nil, and the problem which their parents to-day present will repeat itself in the course of years.

They would need more supervision than was practicable – in consequence they should be 'regarded as a passing phase which would disappear in the course of time'. Stripped of government support to purchase land, these poor whites became the first beneficiaries of the Pact-initiated Old Age Pension Scheme – in effect, the first generation of beneficiaries of the new imperial science of welfare.

Officials focused rather on younger parents and children. 'The all-important aspect of the matter in the opinion of us all,' wrote Welfare Officer Scholtz, 'is the future of the children' (ARB 29 June 1925). Family life was central to the future of children, which was 'the main factor in the solution of the poor white problem [through] correct training under good discipline of the youths of both sexes'. Scholtz argued that by enabling the new generation of youths to 'earn a decent livelihood and become good citizens, they would gradually and permanently remove or overcome the poor white problem'. Boys and girls were therefore *trained* for specific tasks on the plots: boys were taught correct systems of farming on intensive lines, and girls were given 'thorough training in domestic science, poultry keeping, dairying, etc'. Both sexes were encouraged to work for wages in the homes and on the plots of other settlers, where such opportunities arose. This was so that youths would be able to 'assist their parents in running a holding on correct lines as is the case in Denmark and other countries'. Once again officials were deriving their ideas about what they could achieve from another imperial science – that of mass public education.

The officials were, however, divided amongst themselves as to what kind of formal education the children of poor whites required. The dichotomy of views was often between rural and urban perspectives: officials with urban (usually labour movement) credentials often seem to have defended vocational training for poor-white children in rural areas as against an academic curriculum in general. Those from rural backgrounds felt that vocational education sentenced rural children – such as themselves – to a life of economic and social marginalisation whilst future progress lay in the urban and industrial centres. This latter group of officials supported public schools that were based on an academic curriculum; they protested that the extension of public farm schools in the 1920s, in which rural life was idealised and the curriculum and education regime groomed children for a life 'on the land', was anachronistic and retrogressive for poor whites. By the 1930s, a new policy had been introduced of large, centralised schools with boarding facilities in prominent rural towns for white children from rural areas.

Reflections on a visionary project

Secretary Cousins spoke with a true reformer's passion when he was asked by the Pact Cabinet to justify his programme at Losperfontein:

> In terms of money it would be a long time before they could talk of success. But there were other aspects, and there they could claim success. A large number of people were being employed under favourable conditions – and that was a great achievement. They had instilled in these people a sense of discipline – which was by no means a pleasant or easy task to perform. They had made consider-able progress, and had taught the people improved methods of farming – which was all to the good...They had taught them that a small holding worked intelligently and fertilised by the sweat of the brow would produce more than the big areas to which these men had been accustomed. They had demonstrated that those who worked could earn more than the relief worker's wage of 5/6d a day (ACL 20–21 April 1927: Statement by CW Cousins).

Minister Creswell concurred: 'If results were measured only in terms of pounds, shillings and pence and the success met with the rehabilitation efforts were not considered, then they might as well scrap the scheme and sell the farm for juvenile exploitation' (ACL 20–21 April 1927: Statement by Minister Creswell). The South African government did not scrap the scheme, however, and the Hartbeespoort Dam irrigation settlement has stood for many years as a monument to the efforts of these early white South African officials and politicians to create a permanent white presence on the land of the 'Cradle of Humankind' These officials wanted to create a mould for poor whites that would result in the production of a new kind of fully 'modern' colonial citizen. Such citizens would ensure that white supremacy was solidified and entrenched in South Africa's countryside. Their success must be measured in part by the fact that landownership and agricultural production in the Cradle area remain still in the hands of the descendants of these first poor whites who were partly dragged, partly self-propelled into the mod-ern racial order. This is the real story behind Hartbeespoort Dam.

President Thabo Mbeki and Premier Mbhazima Shilowa at the opening
of Maropeng Heritage Centre

VOICE OF POLITICS, VOICE OF SCIENCE: POLITICS AND SCIENCE AFTER 1945

Philip Bonner, Amanda Esterhuysen and Trefor Jenkins

The post-Second World War period witnessed tumultuous change in South Africa. Some of this arose in the realm of palaeontological research. In 1942 Broom had recognised at least part of the antiquity of the Cradle hominids and formations. In 1946, with the support of Smuts, he published his Transvaal Museum memoir which finally convinced the world that Africa represented the birthplace of humankind. In 1947 he exposed the most important hominid specimen to be found since the Taung skull, the fossil hominid that became known as 'Mrs Ples'. South Africa, in palaeontological terms, was on the crest of a wave. Public and official recognition followed, but was soon brought to a precipitate end as both Afrikaner and African nationalism flexed their muscles and came into collision. In 1948 Afrikaner nationalists swept to power; their National Party would stay in office as the country's controlling political force for the next forty-six years. They came armed with the programme of apartheid or full-scale racial separation. Whereas the inter-war generation of South African politicians had subscribed to a policy of a common white South Africanism, which Smuts in particular sought to anchor in the achievements of a white South African science, the Nationalists narrowed their angle of vision to focus on an Afrikaner 'volk' or nation whom they promoted as God's chosen race. Apartheid looked for intellectual credentials elsewhere – in eugenics, in intelligence testing, but above all in what Loubser (1987) and Dubow (1992) term 'the Apartheid Bible'. Afrikaner theologians and intellectuals ransacked the Old and New Testaments to provide an ideological rationalisation for racial separation and white supremacy. At the core of the argument that emerged, as Dubow shows, is the notion of God as 'Hammabdil', the Great Divider, separating light from dark, Heaven from Earth, men from women, nation from nation. The core text of this new theology was that section of the Old Testament that describes the Tower of Babel. Here God intervenes to disperse the builders of the Tower who wish to create a single nation, by causing them to speak mutually incomprehensible languages. On a different continent and in a different context, Afrikaner nationalists were now repeating this act and observing God's will (Dubow 1992). From this point on, science and public political discourse proceeded on two divergent paths.

With the exception of the Potchefstroom University for Christian National Higher Education, Afrikaans-medium tertiary institutions did not expressly prohibit the teaching of evolution. However, they did little to promote it. For example, the University of Stellenbosch Zoology Department was, in the words of Phillip Tobias, 'steeped in palaeontology' and actively taught evolution, yet at the same time it housed the biggest theology college in South Africa. Tobias recalls the contradictory pulls this exerted on the academic community, as was evident when he was on one occasion invited by the Stellenbosch senate to deliver a lecture to the university; it was attended by an audience of which over 50 per cent consisted of theologians and theology students. The University of the Orange Free State, likewise, researched palaeoanthropology, mostly of fossils of a more recent order, and Professor TF Dreyer, who evinced a distinct Afrikaner national-ist antipathy to foreigners who came to South Africa to pronounce on its fossils (for example Dart, Drennan and Broom), was responsible for the discovery of the 300 000-year-old Florisbad fossil. Both the Bloemfontein and Transvaal Museums, which were closely connected to the universities of the Orange Free State and Pretoria respectively, found themselves in a more delicate and exposed position than the universities them-selves. Both continued actively to pursue palaeoanthropological research, the Transvaal Museum continuing to be in the forefront of this field, but this was more tolerated than encouraged. The National Party-dominated Boards of both museums, for example, only permitted their researchers to mount displays on evolution provided that they indicated prominently that creationism was an alternative, and widely supported, biblically-based interpretation of humankind's beginning. Elsewhere more generally, the most influential academics and intellectuals of their day were Geoff Cronjé, a sociologist, Hendrik Verwoerd, a social psychologist, Gerrit Eloff, a eugenicist, Piet Meyer and Nico Diederichs, political philosophers, all of whom were National-Party aligned, while Christian National Education, the pedagogical framework to which the government subscribed and which repudiated theories of evolution, was enforced throughout the primary and secondary educational system.

Christian National Education, however, did more than simply exclude evolution; it developed, reinforced and perpetuated a racist creationist ideology that would have a far greater impact than a mere lack of knowledge about the subject. CNE explicitly set out to Christianise *all* South Africans. The motives for 'christianising' different 'races', how-ever, differed. It was felt that there was an obligation to 'Christianise' the 'black and coloured' child to 'secure him against his own heathen and all kinds of ideologies which promise him sham happiness, but in the long run will make him unsatisfied and un-happy' (Christian National Education Policy 1949: Articles 14 & 15; see also Enslin 1984: 140). Christian national principles were entrenched in white children by means of the religious assemblies, religious instruction, youth preparedness programmes at schools and veld-schools run by the Department of National Education to alert them to the evils of communism, justify patriotism and existing race relations, reinforce white superiority and cultivate an appreciation of the biblical story of Creation (Christie 1991: 185–186).

Notions of 'racial superiority', coupled with creationism, played out as an odd mix of what can best be described as a cross between Victorian polygenism and Darwin's suggestion that some 'races' were vestiges of different stages in the evolution of 'mankind' (see Gould 1981: 36). Many white people came to believe that they had been created, but that black people had evolved, and many black people regarded activities at Sterkfontein with suspicion because they understandably believed that the motive behind the excavation was to prove that the 'blacks' had evolved. Evolution thus not only touched on religious sensitivities but cut to the core of South African identity politics.

Meanwhile, English-medium universities and other institutions retreated into their shells, becoming increasingly isolated and insulated from public life by the government's preoccupation with *volkekunde* (ethnography) and simple exclusion. Many liberal academics took up teaching positions overseas, especially after the Extension of University Education Act 45 of 1959 (a typical euphemism; the Act in effect restricted the access of black students to universities throughout the country), and the Sharpeville massacre in 1960. The remainder looked to Oxford, Harvard or, in Tobias's case, Nairobi, as their point of intellectual reference and affirmation. Research into palaeontology continued – even flourished – in some museums and English-medium universities, as the contents of this book make clear. For example, CK (Bob) Brain at the Transvaal Museum began his 25-year excavation at Swartkrans, one of Broom's earlier 'hunting grounds', in 1964, and Tobias and Hughes embarked on a new programme at Sterkfontein in 1966 which continues to this day, and which was supported by the government-funded Council for Scientific and Industrial Research; but in wider public life this and other disciplines focusing on an indigenous African past were muzzled and pushed to the margins. A curious kind of national split consciousness developed, the intellectual history of which has yet to be written, and which would persist into the 1990s when a new South Africa was born. Then once more, the 'Cradle of Humankind' could come into its own, and again be drawn into the construction of a new national identity.

In 1998 Ron Clarke announced the discovery of the near-complete skeleton of a 3.3–4.0-million-year-old hominid at Sterkfontein. Phillip Tobias and Ron Clarke 'marketed' their find in Parliament, significantly, as the 'African Naissance', and they garnered support from then Deputy President Thabo Mbeki, who attended the press announcement. Shortly after this, in 1999, Sterkfontein, Kromdraai and their environs were granted World Heritage status and were marketed by the government as the 'Cradle of Humankind', the home to all humanity. By making everyone African, President Mbeki situated Africa firmly at the centre of history and provided a moment in time when all South Africans had shared a past:

> I would ask you to be very still. If we are very still, we will hear, if we really listen, these rocks and stones speaking to us today.
>
> They are the voices of our distant ancestors, who still lie buried in them.
>
> The voices of my ancestors and yours!
>
> You see, in Africa, things are seldom what they seem. And so I would say to everyone, welcome home! (Mbeki 2005)

Clearly, scientific advances at the Cradle of Humankind World Heritage Site have benefited from successive phases of official state interest and endorsement. The meaning of the 'Cradle of Humankind' will continue to be excavated scientifically and constructed culturally, as the political imperatives of the country change.

NOTES, REFERENCES
AND RECOMMENDED READING

PART 1
AFRICA IS SELDOM WHAT IT SEEMS

Reference

Dubow S. 1995. *Illicit Union. Scientific racism in modern South Africa.* Johannesburg: Wits University Press

CHAPTER 1
WHITE SOUTH AFRICA AND THE SOUTH AFRICANISATION OF SCIENCE

Notes

1 The present chapter is adapted from Dubow S. 2006. Oxford: Oxford University Press

2 The timing of this debate closely followed that in Britain, where argument about the antiquity of humans moved centre stage in the mid-1860s, as a consequence of research conducted at Brixham cave in 1858 and the publication of work by Lyell and Huxley in 1863. The first mention of shell caves at the Cape was made by way of a 'Query' by 'Barnacle' in the *Cape Monthly Magazine* in June 1858.

3 Dales definition recall's Lubbock's characterisation of the 'new science' of archaeology which 'forms, in fact, the link between geology and history'. See Lubbock J. 1872. *Prehistoric Races as Illustrated by the Ancient Remains and the Manners and Customs of Modern Savages.* 3rd ed., London, p. 2.

4 In 'A Commentary on the History and Present Position' (Goodwin 1928: 295–296), Goodwin notes that Dale's finds were cited as evidence by John Lubbock in his struggle with Thomas Huxley for control of the Anthropological Institute of Great Britain and Ireland.

References

Anker P. 2001. *Imperial Ecology. Environmental Order in the British Empire 1895–1945.* Cambridge, Mass.: Harvard University Press

Burkitt MC. 1928. *South Africa's Past in Stone and Paint.* Cambridge: Cambridge University Press

Dart R. 1925. Australopithecus africanus: The Man-Ape of South Africa. *Nature* 7 February

Delta (Langham Dale). 1870. Stone Implements in South Africa. *Cape Monthly Magazine* October

Diop D. 1999. Africa: Mankind's Past and Future. In MW Makgoba (ed.) *African Renaissance. The New Struggle.* Sandton & Cape Town: Mafube, Tafelberg

Dubow S. 1995. *Scientific Racism in Modern South Africa.* Cambridge: Cambridge University Press

Dubow S. 1996. Human Origins, Race Typology and the Other Raymond Dart. *African Studies* Volume 55 No. 1

Dubow S. 2004. Earth History, Natural History, and Prehistory at the Cape, 1860–1875. *Comparative Studies in Society and History* 46

Dubow S. 2006. *A Commonwealth of Knowledge: Science, Sensibility and White South Africa 1820–2000.* Oxford: Oxford University Press

Dubow S & Marks S. 2001. Patriotism of Place and Race: Keith Hancock on South Africa. In DA Low (ed.) *Keith Hancock. The Legacies of a Historian.* Melbourne: Melbourne University Press

Goodwin AJH. 1928. Sir Langham Dale's Collection of Stone Implements. *South African Journal of Science* XXV

Gordon RJ. 1992. *The Bushman Myth.* Boulder: Westview Press

Moaholi N. 1999. Afrikatourism. In MW Makgoba (ed.) *African Renaissance. The New Struggle* Sandton & Cape Town: Mafube, Tafelberg

Smuts JC. 1925. South Africa in Science. *South African Journal of Science* 22

Stafford RA. 1999. Scientific Exploration and Empire. In A Porter (ed.) *The Oxford History of the British Empire.* Oxford: Oxford University Press

Tobias PV. 2002. Africa: The Cradle of Humanity. In H Bajinath & Y Singh *Rebirth of Science in Africa. A Shared Vision for Life and Environmental Sciences.* Pretoria: Umdaus Press

PART 2
FOSSILS AND GENES

References

Brain CK. (ed). 1958. *The Transvaal Ape-man Bearing Cave Deposits.* Transvaal Museum Memoir No. 11. Transvaal Museum, Pretoria

Brain CK. 2004. Introduction. *Swartkrans: A Cave's Chronicle of Early Man.* Second edition Transvaal Museum Monograph No. 8

Brain CK & Sillen A. 1988. Evidence from the Swartkrans Cave for the Earliest Use of Fire. *Nature* 336: 464–466

Broom R. 1950. *Finding the Missing Link.* London: Watts & Co

Cann R, Stoneking M & Wilson A. 1987. Mitochondrial DNA and Human Evolution. *Nature* 325: 31–36

Chatwin B. 1987. *The Songlines.* London: Jonathan Cape

Grine FE. 2004. Description and Preliminary Analysis of New Hominid Craniodental Fossils from the Swartkrans Formation. In CK Brain (ed.) *Swartkrans: A Cave's Chronicle of Early Man* (2nd ed.). Transvaal Museum Monograph No. 8

Jones S. 1994. *The Language of the Genes.* London: Flamingo

Krings M, Capelli C, Tschentsher F, Geisert H, Meyer S, Von Haeseler A, Grosschmidt K, Possnert G, Paunovic M & Pääbo S. 2000. A View of Neanderthal Genetic Diversity. *Nature Genetics* 26: 144–146

Leakey L, Tobias PV & Napier J. 1964. A New Species of the Genus *Homo* from Olduvai Gorge. *Nature* 202: 7–9

Shakespeare N. 1999. *Bruce Chatwin.* London: The Harvill Press with Jonathan Cape

Soodyall H. (ed.). 2006. *The Prehistory of Africa: Tracing the Lineage of Modern Man.* Johannesburg: Jonathan Ball Publishers

Stringer C. 2006. The origins of modern humans 1984–2004. In Soodyall H (ed.). *The Prehistory of Africa: Tracing the Lineage of Modern Man.* Johannesburg: Jonathan Ball Publishers

Stringer C & Andrews P. 2005. *The Complete World of Human Evolution.* London: Thames and Hudson

Recommended reading

Broom R. 1936. A New Fossil Anthropoid Skull from South Africa. *Nature* 138: 486–488

Clarke RJ. 1998. First ever Discovery of a Well-preserved Skull and Associated Skeleton of *Australopithecus. South African Journal of Science* 94: 460–463

Dart RA. 1925. *Australopithecus africanus,* the Man-ape of South Africa. *Nature* 115: 195–199

CHAPTER 2
A HISTORY OF SOUTH AFRICAN PALAEOANTHROPOLOGY

Notes

1 Palaeoanthropology, spelled paleoanthropology in the USA, is derived from three Greek words: *palaios* – ancient, *anthrōpos* – human and *logos* – word, reason.

2 Anthropology is a scientific discipline that studies humans. There is one major subdivision of the discipline: if it studies social or cultural aspects of humanity it is called social or cultural anthropology; if the emphasis is on the biological side it is known as biological (physical) anthropology. Palaeoanthropology is one of the sub-disciplines of biological anthropology. Broca's programme was ambitious and the society soon established educational and research facilities. Anthropological societies appeared in other countries soon afterwards.

3 *Hominids* are a group of primates that includes humans and all their ancestors and closely related forms. Now often referred to as *hominins,* this group includes the genus *Homo,* australopithecines and earlier species, but not chimpanzees, gorillas or their ancestors.

4 *Phylogeny* refers to the evolutionary relationships among species through time, and is commonly depicted as an 'evolutionary tree'.

5 The Cradle of Humankind in South Africa was established in 1999 as a UNESCO World Heritage Site (WHS). It includes numerous archaeological and fossil localities in the region near Sterkfontein, Swartkrans and Kromdraai. In 2005, the other major South African Plio-Pleistocene sites (dating from 7.00 to 0.05 million years ago) at Taung and the Makapansgat Valley were granted status as WHS extension sites to the Cradle.

6 One of the main proponents of the idea of cultural diffusion was the Australian neuroanatomist Sir Grafton Elliot Smith, who explained both cultural and physical similarities among ancient populations as the consequence of a process of diffusion (essentially, migration) from hypothesised centres of origin in Europe or Asia. Diffusion was also used to explain the presence of both 'primitive' and 'advanced' physical characteristics observed on fossil crania all over the world, and individual crania were taken to represent ancient 'types' at different evolutionary stages of development.

7 See Dart & Craig (1959) *Adventures With the Missing Link*; Tobias (1984) *Dart, Taung and the Missing Link.*

8 The ODK was an elaborate cultural theory involving australopithecine modification and use of tools made from animal bones (osteo-), teeth (-donto-), and horns (-keratic). Dart based this theory mostly on his description and interpretation of animal bone 'tools' from Makapansgat, and published thirty-nine papers between 1949 and 1965. The theory has since been discredited, but is recognised as the stimulus for the first naturalistic taphonomic studies (i.e. studies of burial processes) in palaeoanthropology.

9 The nomenclature used to accession fossils into a collection usually involves a site-name abbreviation (e.g. Sts or StW = Sterkfontein, SK = Swartkrans), followed by a sequential catalogue number.

10 The new genus name *Paranthropus* translates as 'next to man', reflecting Broom's interpretation that they differed sufficiently from the Sterkfontein material to belong in a separate lineage from that leading to modern humans.

References

Bowler PJ. 1986. *Theories of Human Evolution: A Century of Debate, 1844–1944.* Baltimore: Johns Hopkins University Press: Dart & Craig 1959

Recommended reading

Delisle RG. 2006. *Debating Humankind's Place in Nature 1860–2000: The Nature of paleoanthropology.* Upper Saddle River: Pearson Prentice Hall

Findlay GH. 1972. *Dr Robert Broom, FRS: Palaeontologist and Physician / 1866–1951.* Cape Town: AA Balkema

Gundling T. 2005. *First in Line: Tracing Our Ape Ancestry.* New Haven & London: Yale University Press

Lewin R. 1997. *Bones of Contention: Controversies in the Search for Human Origins.* Second edition. Chicago: University of Chicago Press

Reader J. 1988. *Missing Links: The Hunt for Earliest Man.* Harmondsworth: Penguin

Štraklj G. 2003. Robert Broom's Theory of Evolution. *Transactions of the Royal Society of South Africa* 58(1): 35–39

Tattersall I. 1995. *The Fossil Trail: How We Know What We Think We Know About Human Evolution.* New York & Oxford: Oxford University Press

Tobias PV. 1984. *Dart, Taung and the 'Missing Link'*. Johannesburg: Witwatersrand University Press

Tobias PV. 1985. History of Physical Anthropology in South Africa. *Yearbook of Physical Anthropology* 28: 1–52

CHAPTER 3

FOSSIL HOMINIDS OF THE 'CRADLE OF HUMANKIND'

Note

1 A niche is a term in ecology for the biological role of an organism, and includes the kind of habitat in which the organism exists, as well as behavioural factors such as mode of locomotion, diet and activity periods (e.g., day-time or nocturnal).

References

Asfaw B, White T, Lovejoy O, Latimer B, Simpson S & Suwa G. 1999. Australopithecus garhi: A New Species of Eearly Hominid from Ethiopia. *Science* 284: 629–635

Brain CK. 1981. *The Hunters or the Hunted? An Introduction to African Cave Taphonomy*. Chicago: University of Chicago Press

Clarke RJ. 1998. First Ever Discovery of a Well-preserved Skull and Associated Skeleton of *Australopithecus. South African Journal of Science* 94: 460–463

Curnoe D. 2006. Early *Homo* in Southern Africa (abstract). African Genesis: A symposium on hominid evolution in Africa. University of the Witwatersrand Medical School, 8–14 January 2006. *Abstracts and Information* volume. Johannesburg: Content Solutions

Curnoe D & Tobias PV. 2006. Description, New Reconstruction, Comparative Anatomy, and Classification of the Sterkfontein Stw 53 Cranium, with Discussions about the Taxonomy of Other Southern African Early *Homo* Remains. *Journal of Human Evolution* 50: 36–77

Dart RA & Craig D. 1959. *Adventures with the Missing Link*. New York: The Viking Press

Dennell R. 2003. Dispersal and Colonisation, Long and Short Chronologies: How Continuous is the Early Pleistocene Record for Hominids Outside East Africa? *Journal of Human Evolution* 45: 421–440

Dennell R & Roebroeks W. 2005. An Asian perspective on Early Human Dispersal from Africa. *Nature* 438: 1099–1104

Recommended reading

Johanson D & Edgar B. 1996. *From Lucy to Language*. London: Weidenfeld & Nicolson

Keyser AW. 2000. The Drimolen Skull: The Most Complete Australopithecine Cranium and Mandible to Date. *South African Journal of Science* 96: 189–193

McHenry HM & Berger LR. 1998. Body Proportions in *Australopithecus afarensis* and *A. africanus* and the Origin of the Genus *Homo. Journal of Human Evolution* 35: 1–22

Rak Y. 1983. *The Australopithecine Face*. New York: Academic Press

Thackeray JF. 1997. Cranial Bone of 'Mrs Ples' (Sts 5): Fragments Adhering to Matrix. *South African Journal of Science* 93: 169–170

Thackeray JF, Braga J, Triel N & Labuschagne JH. 2002. 'Mrs Ples' (Sts 5) from Sterkfontein: An Adolescent Male? *South African Journal of Science* 98: 21–22

CHAPTER 4

THE CONTRIBUTION OF GENETIC STUDIES

References

Barkhan D. 2003. Haploid Genetic Variation in Populations from Uganda, Zambia and the Central African Republic. PhD thesis, University of the Witwatersrand

Brain P. 1952a. The Sickle-cell Trait: Its Clinical Significance. *South African Medical Journal* 26: 925–928

Brain P. 1952b. Sickle-cell Anaemia in Africa.

British Medical Journal ii: 880

Brain P. 1953. The Sickle-cell Trait: A Possible Mode of Introduction into Africa. *Man* 233

Brain P. 1956. The Sickle-cell Phenomenon. *Central African Journal of Medicine* 2: 73–77

Cann RL, Stoneking M & Wilson AC. 1987. Mitochondrial DNA and Human Eevolution. *Nature* 325: 31–36

Chen Y-S, Olckers A, Schurr TG, Kogelnik AM, Huoponen K & Wallace DC. 2000. mtDNA Variation in the South African Kung and Khwe – And Their Genetic Relationships to Other African Populations. *American Journal of Human Genetics* 66: 1362–1383

Clark AG, Glanowski S, Nielsen R, Thomas PD, Kejariwal A, Todd MA, Tanenbaum DM, Civello D, Lu F, Murphy B, Ferreira S, Wang G, Zheng X, White TJ, Sninsky JJ, Adams MD & Cargill M. 2003. Inferring Nonneutral Evolution from Human-Chimp-Mouse Orthologous Gene Trios. *Science* 302: 1960–1963

Cruciani F, La Fratta R, Santolamazza P, Sellitto D, Pascone R, Moral P, Watson E, Guida V, Colomb EB, Zaharova B, Lavinha J, Vona G, Aman R, Cali F, Akar N, Richards M, Torroni A, Novelletto A & Scozzari R. 2004. Phylogenetic Analysis of Haplogroup E3b (E-M215) Y Chromosomes Reveals Multiple Migratory Events Within and Out of Africa. *American Journal of Human Genetics* 74: 1014–1022

Currie P. 2004. Muscling in on Hominid Evolution. *News and Views, Science* 428: 373–374

Dart RA. 1937a. The Hut Distribution, Genealogy and Homogeneity of the /?Aauni-≠Khomani Bushmen. *Bantu Studies* ii: 159–174

Dart RA. 1937b. The physical characters of the /?Auni-≠Khomani Bushmen. *Bantu Studies* ii: 175–246

Dart RA. 1951. African Serological Patterns and Human Migrations. Paper presented at the South African Archaeological Society Meeting, Cape Town, South Africa

Elsdon-Dew R. 1936. The Blood Groups of the Bantu of Southern Africa. *Publications of the South African Institute for Medical Research* 7: 217–300

Elsdon-Dew R. 1939. Blood Groups in Africa. *Publications of the South African Institute for Medical Research* 9: 29–94

Enard W, Przeworski M, Fisher SE, Lai CS, Wiebe V, Kitano T, Monaco AP Paabo S. 2002. Intra- and Interspecific Variation in Primate Gene Expression Patterns. *Science* 296: 340–343

Forster P. 2004. Ice Ages and the Mitochondrial DNA Chronology of Human Dispersals: A Review. *Philosophical Transactions of the Royal Society of London* 359B: 255–264

Hirschfeld L & Hirschfeld HI. 1919. Serological Differences between the Blood of Different Races: The result of Researches on the Macedonian Front. *Lancet* ii: 675–679

Ingman M, Kaessmann H, Pääbo S & Gyllensten U. 2000. Mitochondrial Genome Variation and the Origins of Modern Humans. *Nature* 408: 708–713

Jenkins T. 1963. Sickle Cell Anaemia in Wankie, Southern Rhodesia. *Central African Journal of Medicine* 8: 212–216

Jenkins T. 1965. Ability to Taste Phenylthio-carbamide in Kalahari Bushmen and Southern Bantu. *Human Biology* 37: 371–374

Jenkins T, Blecher SR, Smith AN & Anderson CG. 1968. Some Hereditary Red-cell Traits in Kalahari Bushmen and Bantu: Hemoglobins, Glucose-6-phosphate Dehydrogenase Deficiency and Blood Groups. *American Journal of Human Genetics* 20: 229–309

Jenkins T, Zoutendyk A & Steinberg AG. 1970. Gamma Globulin Groups (Gm and Inv) of Various Southern African Populations. *American Journal of Physical Anthropology* 32: 197–218

Krings M, Capelli C, Tschentscher F, Geisert H, Meyer S, von Haeseler A, Grosschmidt K,

Possnert G, Paunovic M & Pääbo S. 2000. A View of Neandertal Genetic Diversity. *Nature Genetics* 26: 144–146

Krings M, Stone A, Schmitz RW, Krainitzki H, Stoneking M. & Pääbo S. 1997. Neanderthal DNA Sequences and the Origin of Modern Humans. *Cell* 90: 19–30

Lane AB, Soodyall H, Arndt S, Ratshikhopha ME, Jonker E, Freeman C, Young L, Morar B & Toffie L. 2002. Genetic Substructure in South African Bantu-speakers: Evidence from Autosomal DNA and Y Chromosome Studies. *American Journal of Physical Anthropology* 119: 175–185

McDonald, D. 2005. Y Haplogroups of the World. www.scs.uiuc.edu/mcdonald/WorldHaplogroupsMaps.pdE

Mourant AE. 1961. Evolution, Genetics and Anthropology. *Journal of the Royal Anthropological Institute* 91: 151–165

Nurse G, Weiner JS & Jenkins T. 1985. *The Peoples of Southern Africa and their Affinities*. Oxford: Clarendon Press

Ovchinnikov IV, Götherström A, Romanova P, Kharitonov VM, Lidén K & Goodwin W. 2000. Molecular Analysis of Neanderthal DNA from the Northern Caucasus. *Nature* 404: 490–493

Pijper A. 1929–30. The Blood Groups of the Bantu. *Transactions of the Royal Society of South Africa* 18: 311–315

Pijper A. 1932. Blood Groups of Bushmen. *South African Medical Journal* 6: 35–37

Pijper A. 1935. Blood Groups in Hottentots. *South African Medical Journal* 9: 192–195

Pirie JHH. 1921. Blood Testing Preliminary to Transfusion, with a Note on the Group Distribution among SA Natives. *Medical Journal of South Africa* 16: 109–112

Salas A, Richards M, De la Fe T, Lareu M, Sobrino B, Sanchez-Diz P, Macaulay V & Carracedo A. 2002. The Making of the African mtDNA Landscape. *American Journal of Human Genetics* 71: 1082–1111

Shapiro M. 1951. The ABO, MN, P and Rh Blood Group Systems in the South African Bantu. *South African Medical Journal* 25: 165–170; 187–192

Soodyall H. 1993. Mitochondrial DNA Variation in Southern African Populations. PhD thesis, University of the Witwatersrand

Singer R. 1958. The Boskop 'Race' Problem. *Man* 232: 173–178

Spurdle AB & Jenkins T. 1992. The Y Chromosome as a Tool for Studying Human Evolution. *Current Opinion and Genetic Development* 3: 487–491

Stedman HH, Kozyak BW, Nelson A, Thesier DM, Su LT, Low DW, Bridges CR, Shrager JB, Minugh-Purvis N & Mitchell MA. 2004. Myosin Gene Mutation Correlates with Anatomical Changes in the Human Lineage. *Nature* 428: 415–418

Stringer C. 2002. Modern Human Origins: Progress and Prospects. *Philosophical Transactions of the Royal Society of London* 357B: 563–579

Underhill PA, Passarino G, Lin AA, Shen P, Mirazon Lahr M, Foley RA, Oefner PJ & Cavalli-Sforza LL. 2001. The Phylogeography of Y Chromosome Binary Haplotypes and the Origins of Modern Human Populations. *Annals of Human Genetics* 65: 43–62

Underhill PA, Shen P, Lin AA, Jin L, Passarino G, Yang WH, Kauffman E, Bonné-Tamir B, Bertranpetit J, Francalacci P, Ibrahim M, Jenkins T, Kidd JR, Mehdi SQ, Seielstad MT, Wells RS, Piazza A, Davis RW, Feldman MW, Cavalli-Sforza LL & Oefner PJ. 2000. Y Chromosome Sequence Variation and the History of Human Populations. *Nature Genetics* 26: 358–361

Watson E, Brauer K, Aman R, Weiss G, von Haeseler A & Pääbo S. 1996. mtDNA Sequence Diversity in Africa. *American Journal of Human Genetics* 59: 437–444

Wolpoff MH, Hawks J & Caspari R. 2000.

Multiregional, not Multiple Origins. *American Journal of Physical Anthropology* 112: 129–136

Zoutendyk A, Kopec AC & Mourant AE. 1953. The Blood Groups of the Bushman. *American Journal of Physical Anthropology* 11: 361–368

Zoutendyk A, Kopec AC & Mourant AE. 1955. The Blood Groups of the Hottentots. *American Journal of Physical Anthropology* 13: 691–698

CHAPTER 5

FOSSIL PLANTS FROM THE 'CRADLE OF HUMANKIND'

Note

1 I thank PAST (Palaeoanthropological Scientific Trust) for ongoing financial support, the Leakey Foundation for a grant to visit the modern wood collection at the Musée Royal de l'Afrique Centrale, Belgium, and the Bernard Price Institute for Palaeontological Research for technical support.

References

Avery DM. 2001. The Plio-Pleistocene Vegetation and Climate of Sterkfontein and Swartkrans, South Africa, Based on Micromammals. *Journal of Human Evolution* 41: 113–132

Bamford MK. 1999. Pliocene Fossil Woods from an Early Hominid Cave Deposit, Sterkfontein, South Africa. *South African Journal of Science* 95: 231–237

Berger LR & Tobias PV. 1994. New Discoveries at the Early Hominid Site of Gladysvale, South Africa. *South African Journal of Science* 90: 223–226

Cadman A & Rayner RJ. 1989. Climatic Change and the Appearance of *Australopithecus africanus* in the Makapansgat Sediments. *Journal of Human Evolution* 18: 107–113

Carrión JS & Scott L. 1999. The Challenge of Pollen Analysis in Palaeoenvironmental Studies of Hominid Beds: The Record from Sterkfontein caves. *Journal of Human Evolution* 36: 401–408

Efremov IA. 1940. Taphonomy: A New Branch of Palaeontology. *Pan American Geologist* 74: 81–93

Horowitz A. 1975. Preliminary Palaeoenvironmental Implication of Pollen Analysis of Middle Breccia from Sterkfontein. *Nature* 858: 417–418

Horowitz A. 1992. *Palynology of Arid Lands*. Amsterdam: Elsevier

Martín-Closas C & Gomez G. 2004. Taphonomie des Plantes et Interprétations Paléoécologiques. Une Synthèse. *Géobios* 37: 65–88

Palmer E & Pitman N. 1972. *Trees of Southern Africa* Vol. III. Cape Town: AA Balkema

Prance GT. 1972. A Monograph of the Neotropical Dichapetalaceae. *Flora Neotropica* Monograph 10: 3–84

Reed KE. 1997. Early Hominid Evolution and Ecological Change through the African Plio-Pleistocene. *Journal of Human Evolution* 32: 289–322

Ruddiman WF. 2004. Early Anthropogenic Overprints on Holocene Climate. *PAGES News* 12(1): 18–19

Schwab MJ, Neumann F, Litt T, Negendank JFW & Stein M. 2004. Holocene Palaeoecology of the Golan Heights (Near East): Investigation of Lacustrine Sediments from Birkat Ram Crater Lake. *Quaternary Science Reviews* 23: 1723–1731

Scott L. 1982. Pollen Analyses of Late Cainozoic Deposits in the Transvaal, South Africa, and their Bearing on Palaeoclimates. *Palaeoecology of Africa* 15: 101–107

Scott L & Bonnefille R. 1986. Search for Pollen from the Hominid Deposits of Kromdraai, Sterkfontein and Swartkrans: Some Problems and Preliminary Results. *South African Journal of Science* 82: 380–382

Zavada MS & Cadman A. 1993. Palynological Investigations at the Makapansgat Limeworks: An Australopithecine Site. *Journal of Human Evolution* 25: 337–350

PART 3

THE EMERGING STONE AGE

References

Brain CK. 1981. *The Hunters or the Hunted? An Introduction to African Cave Taphonomy.* Chicago: University of Chicago Press

Dart RA. 1956. Cultural Status of the South African Man-apes. *Smithsonian Report* No. 4240: 317–338

Deacon J. 1990. Weaving the Fabric of Stone Age Research in Southern Africa. In P. Robert Shaw (ed.) *A History of African Archaeology*. London: James Currey

Federasie van Afrikaanse Kultuurverenigings. 1948. *Instituut vir Christelike-Nationale Onderwys, Beleid.* Johannesburg

Forde CD. 1929. South Africa's Past in Stone. *American Anthropologist* New Series Vol. 31(1): 156–163

Gould SJ. 1992. *The Mismeasure of Man*. London: Penguin Books

Howell FC. 1966. *Early Man.* Nederland: Time-Life International

Maguire M. 1997. A History of Middle and Late Stone Age Research in the Kwazulu-Natal Coastal Area from 1871 to the Beginning of the Second World War. In JA van Schalkwyk, CJ van Vuuren & I Plug. Studies in Honour of Professor JF Eloff. *Research by the National Cultural History Museum* 6: 43–58

Proctor RN. 2003. Three Roots of Human Recency. Molecular Anthropology, the Refigured Achulean, and the UNESCO Response to Auschwitz. *Current Anthropology* 44(2): 229

Rose B & Tunmer R. (eds). 1975. *Documents in South African Education.* Johannesburg: AD Donker

Schlanger N. 2002. Making the Past for South Africa's Future: The Prehistory of Field-Marshal Smuts (1920–1940s). In *Ancestral Archives, Explorations in the History of Archaeology. Antiquity* 76: 200–209

Schlanger N. 2003. The Burkitt Affair Revisited. Colonial Implications and Identity Politics in Early South African Prehistoric Research. *Archaeological Dialogues* 10(1): 5–26

Shepard N. 2003. State of the Discipline: Science, Culture and Identity in South African Archaeology, 1870–2003. *Journal of Southern African Studies* 29(4): 823–844

Trigger BG. 1989. *A History of Archaeological Thought.* Cambridge: Cambridge University Press

Van Hoepen ECN. 1930. Mr C Van Riet Lowe and South African Prehistoric Archaeology. *The South African Journal of Natural History* 6(5): 345–368

Van Riet Lowe C. 1929. 'Die Koningse Kultuur'. *The South African Journal of Natural History* 5(4): 331–334

CHAPTER 6

THE EARLIER STONE AGE

References

Leakey M. 1984. *Disclosing the Past. An Autobiography.* London: Weidenfeld and Nicolson

Recommended reading

Brain CK. 1981. *The Hunters or the Hunted? An Introduction to African Cave Taphonomy.* Chicago: University of Chicago Press

Brain CK. 2002. Interview with Amanda Esterhuysen, filmed by WITS TV

D'Errico F, Backwell LR & Berger L. 2001. Bone Tool Use in Termite Foraging by Early hominids and its Impact on Understanding of Early Hominid Behaviour. *South African Journal of Science* 9(3/4) : 71–75

Kuman K. 2005. Interview with Amanda Esterhuysen

Kuman K & Clarke RJ. 2000a. Stratigraphy, Artefacts Industries and Hominid Associations for Sterkfontein Member 5. *Journal of Human Evolution* 38: 827–847

Kuman K & Clarke RJ. 2000b. *The Sterkfontein*

Caves Palaeontological and Archaeological Site.
Guidebook published by K Kuman & RJ Clarke

CHAPTER 7
THE MIDDLE STONE AGE AND LATER
STONE AGE

References
Deacon HJ & Deacon J. 1999. *Human Beginnings in South Africa: Uncovering the Secrets of the Stone Age.* Cape Town: David Philip
Mitchell PJ. 2002. *The Archaeology of Southern Africa.* Cambridge: Cambridge University Press

CHAPTER 8
ROCK ENGRAVINGS IN THE MAGALIESBERG
VALLEY

Recommended reading
Dowson TA. 1992. *Rock Engravings of Southern Africa.* Johannesburg: Witwatersrand University Press
Lewis-Williams JD & Pearce DG. 2004. *San Spirituality: Roots, Expression, and Social Consequences.* Cape Town: Double Storey
Ouzman S. 1996. Thaba Sione: Place of Rhinoceroses and Rock Art. *African Studies* 55: 31–59
Ouzman S. 2001. Seeing is Deceiving: Rock Art and the Non-Visual. *World Archaeology* 33(2): 237–256

PART 4
THE MYTH OF THE VACANT LAND

References
Cobbing J. 1984. The Case Against the Mfecane. Unpublished paper presented to the African Studies Seminar 5 March, University of the Witwatersrand
Huffman TN. 1970. The Early Iron Age and the Spread of the Bantu. *South African Archaeological Bulletin* Vol. XXV (Part I) No. 97: 1–21
Inskeep RR. 1970. The archaeological background.

In M Wilson & L Thompson (eds.) *The Oxford History of South Africa* Vol. I. Oxford: Clarendon Press
Krantz ME & Trengrove WE. n.d. *The People of South Africa.* Second edition. Cape Town: n.p.
Mason R. 1973. First Iron Age Settlement in South Africa: Broederstroom 24/73. *South African Journal of Science.* November
Smit GJJ, Kreuser FOA & Vlok AC. 1980. *History for Standard Eight.* Third edition. Cape Town: Maskew Miller Longman
Theal GM. 1903. *History of South Africa, 1794–1828.* London: George Allen and Unwin
Walker E. 1922. *Historical Atlas of South Africa* (map entitled 'The Bantu Devastation and the Great Trek, 1820–1848). Oxford: Oxford University Press

CHAPTER 9
THE EARLY IRON AGE AT BROEDERSTROOM

References
Gramley RM. 1978. Expansion of Bantu-speakers Versus Development of Bantu Language and African Culture in Situ: An Archaeologist's Perspective. *South African Archaeological Bulletin* 33: 107–112
Greenburg JH. 1955. *Studies in African Linguistic Classification.* New Haven: Yale University Press
Hall M. 1986. The Role of Cattle in Southern African Agropastoral Societies: More than Bones Alone can Tell. In M Hall & AB Smith (eds) *Prehistoric Pastoralism in Southern Africa.* South African Archaeological Society Goodwin Series 5: 83–87
Hall M. 1987. *The Changing Past: Farmers, Kings and Traders in Southern Africa.* Cape Town: David Philip
Hammond-Tooke DW. 1984. In Search of the Lineage: The Cape Nguni Case. *Man* 19(1): 77–93
Huffman TN. 1989. *Iron Age Migrations: The Ceramic Sequence in Southern Zambia, Excavations at Gundu and Ndonde.*

Johannesburg: Witwatersrand University Press

Huffman TN. 1990. Broederstroom and the Origins of Cattle-keeping in Southern Africa. *African Studies* 49(2): 1–12

Huffman TN. 1993. Broederstroom and the Central Cattle Pattern. *South African Journal of Science* 89: 220–226

Kuper A. 1982. *Wives for Cattle: Bridewealth and Marriage in Southern Africa*. London: Routledge & Kegan Paul

Maggs TM & Ward V. 1984. Early Iron Age Sites in the Muden Area of Natal. *Annals of the Natal Museum* 26: 105–140

Mason RJ. 1981. Early Iron Age Settlement at Broederstroom 24/73 Transvaal, South Africa. *South African Journal of Science* 77: 401–416

Mason RJ. 1986. *Origins of Black People of Johannesburg and the Southern Western Central Transvaal AD 350–1880*. Occasional Paper 16. Johannesburg: University of the Witwatersrand Archaeological Research Unit

Plug I & Voigt E. 1985. Archaeozoological Studies of Iron Age Communities in Southern Africa. In F Wendorf (ed.) *Advances in World Archaeology* 4: 189–238

Posnansky M. 1961. Iron Age in East and Central Africa – Points of Comparison. *South African Archaeological Bulletin* 16: 134–136

CHAPTER 10

TSWANA HISTORY IN THE BANKENVELD

Notes

1 Readers will be familiar with the name Tshwane from the debates over the renaming of Pretoria. This naming debate nicely underlines the inadequacy of the term 'prehistory', and illustrates the continuing resonance of local pre-colonial events and the ongoing redefinition of their historically-based significance in the present.

2 The Magaliesberg Mountains were also known in the records as the Cashane Mountains. 'Cashane' was the early European travellers' rendition of Kgasoane.

3 Once again Boer trekkers did not render Tswana names correctly, and Maseloane was known as Selon. The farm upon which Molokwane is located is called 'Selons Kraal', and the river nearby is the Selons River.

References

Breutz PL. 1958. Stone Kraals and Stone Hut Villages in South Africa. *Bantu Education Journal*, April

Breutz PL. 1987. *The Batswana and the Origins of Bophuthatswana*. Margate, South Africa: Thumbprint

Lye WF. (ed.). 1975. *Andrew Smith's Journal of his Expedition into the Interior of South Africa. 1834–1836*. Cape Town: AA Balkema, for the South African Museum

Wallis JPR. (ed.). 1945. *The Matabele Journals of Robert Moffat 1829–1860*. Volume 1. London: Chatto & Windus

CHAPTER 11

THE EARLY BOER REPUBLICS

References

Allen V. 1979. *Lady Trader: A Biography of Mrs Sarah Heckford*. London: Collins

Anderson AA. 1974. *Twenty-Five Years in a Waggon*. repr. Cape Town: Struik

Barrow J. 1801–1804. *An Account of Travels into the Interior of South Africa*. Volumes 1 & 2. London: Cadell and Davies

Boeyens JCA. 2003. The Late Iron Age Sequence in the Marico and Early Tswana History. *South African Archaeological Bulletin* 58 (178)

Bradlow F & Bradlow E. (eds). 1979. *William Somerville's Narrative of His Journeys to the Eastern Cape Frontier and to Lattakoe, 1799–1802*. Cape Town: Van Riebeeck Society

Breutz P-L. 1953. *The Tribes of Rustenburg and Pilansberg Districts*. Ethnological Publications No. 28. Pretoria: Department of Native Affairs

Brooke Simons P. 1998. *The Life and Work of Charles Bell*. Cape Town: Fernwood Press

Cachet FL. 1882. *De Worstelstryd der Transvalers aan het Volk van Nederland Verhaald.* Amsterdam: JH Kruyt

Carruthers J. 1995. *Game Protection in the Transvaal 1846 to 1926.* Pretoria: State Archives Service

Chapman J. 1971. *Travels in the Interior of South Africa, 1849–1863.* Volumes 1 & 2. Cape Town: AA Balkema

Child D. 1979. *A Merchant Family in Early Natal.* Cape Town: Balkema

Coetzee JH. (ed.). 1974. *Boere en Jagters in Ou Marico: Vertellinge uit die Geskiedenis van die ou Maricostreek deur C.F. Gronum.* Potchefstroom: Pro Rege

Craig A, Hummel A & Hummel C. (eds). 1994. *Johan August Wahlberg, Travel Journals (and Some Letters) South Africa and Namibia/Botswana, 1838–1856.* Cape Town: Van Riebeeck Society

Delegorgue A. 1997. *Adulphe Delegorgue's Travels in Southern Africa.* Volume 2. Translated by Fleur Webb, introduced and annotated by Stephanie J Alexander and Bill Guest. Durban & Pietermaritzburg: Killie Campbell Africana Library and University of Natal Press

Etherington N. 2001. *The Great Treks: The Transformation of Southern Africa, 1815–1854.* London: Longman

Gassiott HS. 1852. Notes from a Journal kept during a Hunting Tour of South Africa. *Journal of the Royal Geographical Society* 22

Guelke L. 1989. Freehold Farmers and Frontier Settlers, 1657–1780. In R Elphick & H Giliomee (eds) *The Shaping of South African Society, 1652–1840.* Cape Town: Maskew Miller Longman

Harris WC. 1840. *Portraits of the Game and Wild Animals of Southern Africa.* London: n.p.

Harris WC. 1852. *The Wild Sports of Southern Africa.* London: Henry Bohm

Hofmeyr S. 1890. *Twintig Jaren in Zoutpansberg.* Cape Town: JH Rose

Huet P. 1869. *Het Lot der Zwarten in Transvaal.*
Utrecht: Van Peursen

Kennedy RF. (ed.). 1964. *Journal of Residence in Africa, 1842–1853 by Thomas Baines Volume 2, 1850–1853.* Cape Town: Van Riebeeck Society

Kruger P. 1902. *The Memoirs of Paul Kruger.* Volumes 1 & 2. London: T Fisher Unwin

Lye WF. (ed.). 1975. *Andrew Smith's Journal of his Expedition into the Interior of South Africa, 1834–1836.* Cape Town: AA Balkema

McGill DC. 1943. A History of the Transvaal (1853–1864). DPhil thesis, University of Cape Town

Nathan M. 1944. *Paul Kruger: His Life and Times.* Durban: Knox

Parsons N. 1995. Prelude to *Difaqane* in the Interior. In C Hamilton (ed.) *The Mfecane Aftermath.* Johannesburg: Witwatersrand University Press

Pelser AN. 1950. *Geskiedenis van die Suid-Afrikaanse Republic. Deel 1, Wordingsjare.* Cape Town: AA Balkema

Penn N. 2005. *The Forgotten Frontier: Colonist and Khoisan on the Cape's Northern Frontier in the 18th Century.* Athens, Ohio: Ohio University Press

Perry JG. 1931. The Social and Economic Conditions of the Transvaal from 1852 to the first Annexation in 1877. MA dissertation, Rhodes University

Potgieter FJ. 1958. *Die Vestiging van Blankes in Transvaal (1837–1886) met Spesiale Verwysing na die Verhouding Tussen die Mens en die Omgewing.* Pretoria: State Archives Service

Sanderson J. 1981. Memoranda of a Trading Trip into the Orange River (Sovereignty) Free State and the Country of the Transvaal Boers. Originally published in the *Journal of the Royal Geographical Society* 30 (1860). Repr. Pretoria: State Library

Simpson GN. 1986. Peasants and Politics in the Western Transvaal, 1920-1940. MA dissertation, University of the Witwatersrand

Smith A. 1849. *Illustrations of the Zoology of South*

Africa. London: Smith, Elder

Strydom S. 1955. 'n Ondersoek na die Plek- en Plaasname van die Groot Moot. DLitt. thesis, University of Pretoria

Transvaal Archives TA SS8 893/55. Verkoping Hekpoort

Transvaal Archives TAB SS0 R1008/56. GJ Kruger to MW Pretorius

Transvaal Archives TAB A17. Churchill Accession

Van der Merwe PJ. 1938. *Die Trekboer in die Geskiedenis van die Kaapkolonie*. Cape Town: Nasionale Pers

Van der Merwe PJ. 1945. *Trek*. Cape Town: Nasionale Pers

Wagner R. 1983. Zoutpansberg: The Dynamic of a Hunting Frontier. In S Marks & A Atmore (eds) *Economy and Society in Pre-Industrial South Africa*. New York: Longman

Wallis JPR. (ed.). 1946. *The Northern Goldfields Diaries of Thomas Baines*. Volumes 1, 2 & 3. London: Chatto & Windus

Recommended reading

Carruthers V. 2000. *The Magaliesberg*. Johannesburg: Protea Book House

PART 5
THE RACIAL PARADOX

References

Beinart W & Coates P. 1995. *Environment and History*. London: Routledge

Berger D. 1982. White Poverty and Government Policy in South Africa 1892–1934. PhD thesis, Temple University

Dubow S. 1995. *Illicit Union. Scientific racism in Modern South Africa*. Johannesburg: Wits University Press

Schlanger N. 2002. Making the Past for South Africa's Future: The Pre-History of Field-Marshal Smuts (1920s–1940s). *Ancestral Archives. Explorations in the History of Geography, Antiquity* Vol. 76

CHAPTER 12
THE LEGACY OF GOLD

References

Coates PNA. 1987. Pieter Jacob Marais' Search for Gold in the Transvaal. *Contree* 22: 31–33

Malan BD. 1959. Early References to the Sterkfontein Caves, Krugersdorp, South Africa. *South African Journal of Science* 55

Murray BK. 1982. *Wits: The Early Years*. Johannesburg: Wits University Press

Tobias PV. 1983. The Sterkfontein Caves and the Role of the Martinaglia Family. *Adler Museum Bulletin* Special Issue, November.

Wallis JPR. (ed.). 1946. *Northern Goldfields Diaries of Thomas Baines*. Volumes 1–3. London: Chatto & Windus

Recommended reading

Gray J. 1937. *Payable Gold*. Johannesburg: Central News Agency

CHAPTER 13
THE STORY OF STERKFONTEIN SINCE 1895

References

Broom R. 1938a. The Pleistocene Anthropoid Apes of South Africa. *Nature* 142

Broom R. 1938b. Further Evidence on the Structure of the South African Pleistocene Anthropoids. *Nature* 142

Broom R. 1946. The Occurrence and General Structure of the South African Ape-Men. Part 1, *The South African Fossil Ape-men, the Australopithecinae. Transvaal Museum Memoir* 2

Broom R. 1950. *Finding the Missing Link*. London: Watts

Broom R & Schepers GWH. 1946. *The South African Fossil Ape-men, the Australopithecinae. Transvaal Museum Memoir* 2

Dart RA. 1925. *Australopithecus africanus*, the Man-ape of South Africa. *Nature (London)* 115

Dart RA. 1965. *Beyond Antiquity*. Johannesburg: South African Broadcasting Corporation

Dart RA & Craig D. 1959. *Adventures with the Missing Link*. New York: Harper and Brothers

Draper D. 1895. Paper on the Dolomite Rocks of the Bloubank River Valley and on his Visit to the Kromdraai Caves. *Transactions of the South African Geological Society*, 13 May

Draper D. 1897. Discussion of paper on Glacial Theory by Professor Prister. *Transactions of the Geological Society of South Africa* 3

Exton H. 1895. Comment at First Ordinary Meeting of South African Geological Society, 8 April

Exton H. 1898. Presidential Address. *Transactions of the South African Geological Society*, 22 February

Findlay GH. 1972. *D Robert Broom, FRS: Biography, Appreciation and Bibliography*. Cape Town: AA Balkema

Frames ME. 1897. (The animal remains found in the Kromdraai Caves in the Dolomite near Krugersdorp.) *Transactions of the South African Geological Society*

Jones TR. 1937. A New Fossil Primate from Sterkfontein, Krugersdorp, Transvaal. *South African Journal of Science* 33

Malan BD. 1959. Early References to the Sterkfontein Caves, Krugersdorp. *South African Journal of Science* 55

Martinaglia G 1947a. Extract from letter dated 20 August 1947 to the Town Clerk of Krugersdorp, cited in PV Tobias (1983)

Molengraaf GAF. 1904. *Géologie de la Republique Sud-Africaine du Transvaal* (published by the Geological Society of France in 1901, and translated into English by JH Ronaldson in 1904)

Oakley KP. 1960. The History of Sterkfontein, with a Comment by BD Malan. *South African Journal of Science* 56

Silberberg HK. 1979. Letter to PV Tobias, 1 January

Tobias PV. 1979. A Hundred Years of History at Sterkfontein. Annual Commemorative Lecture of the Johannesburg Historical Foundation. 18 September

Tobias PV. 1983. The Sterkfontein Caves and the Role of the Martinaglia Family. *Adler Museum Bulletin* (Festschrift in honour of Dr Cyril Adler on his 80th birthday, November)

Un Frère Mariste. 1898. Les grottes de Sterkfontein. *Cosmos* 47(679)

Van Riet Lowe C. 1947. Die Ontdekking van die Sterkfontein Grotte. *South African Science* 1(4)

Wilkinson MJ. 1973. Sterkfontein Cave System: Evolution of a Karst Form. MA thesis, University of the Witwatersrand

Recommended reading

Martinaglia G 1947b. Extract from letter of May 1947 to C van Riet Lowe, cited in van Riet Lowe (1947)

CHAPTER 14
THE SOUTH AFRICAN WAR 1899–1902

References

Cloete PG. 2000. *The Anglo-Boer War: A Chronology*. Pretoria: Van der Walt

Copley IB. 1993. Ambush at Kalkheuwel Pass, 3 June 1900. *Military History Journal* 9(4)

Hancock WK. 1962. *Smuts Vol. 1. The Sanguine Years 1870–1919*. Cambridge: Cambridge University Press

Joubert DM. 1977. Agricultural Research in South Africa: An Historical Overview. In AC Brown (ed.) *A History of Scientific Endeavour in South Africa*. Cape Town: Royal Society

Smuts JC. 1994. *Memoirs of the Boer War*. Johannesburg: Jonathan Ball

Spies SB. 1977. *Methods of Barbarism? Roberts and Kitchener and Civilians in the Boer War Republics, January 1900 to May 1902*. Cape Town: Human & Rousseau

Taitz J. (ed.). 1996. *The War Memoir of Commandant Ludwig Krause 1899–1900*. Cape Town: Van Riebeeck Society

Warwick P. (ed.). 1980. *The South African War: The Anglo-Boer War 1899–1902*. London: Longman

Warwick P. 1983. *Black People and the South African War 1899–1902*. Johannesburg: Ravan Press

Weber M. n.d. Eighteen Months Under General De la Rey. Unpublished manuscript translated from the German by Juliet Marais Louw

Wulfsohn L. 1987. *Rustenburg at War*. Rustenburg: The author

Recommended reading

Amery LS. 1907. *The Times History of the War in South Africa, 1899–1902 Vol. 5*. London: Sampson Low

Carruthers V. 2000. *The Magaliesberg*. Pretoria: Protea Book House

Grundlingh AM. 1979. *Die Hensoppers and Joiners*. Pretoria: HAUM

Hall DM. 1999. *The Hall Handbook of the Anglo-Boer War 1899–1902*. Pietermaritzburg: University of Natal Press

Meintjes J. 1966. *De la Rey: Lion of the West*. Johannesburg: Hugh Keartland

Nasson B. 1999. *The South African War, 1899–1902*. London: Arnold

Pakenham T. 1979. *The Boer War*. Johannesburg: Jonathan Ball

Pretorius F. 1999. *Life on Commando During the Anglo-Boer War 1899–1902*. Cape Town: Human & Rousseau

Reitz D. 1929. *Commando: A Boer Journal of the Boer War*. London: Faber & Faber

Shaw J. 1999. Dwarsvlei, A Highveld Farm: Forgotten Battlefield of the Anglo-Boer War. *Military History Journal* 11(3/4)

Smith I. 1996. *The Origins of the South African War, 1899–1902*. London: Longman

CHAPTER 15
WHITE SOUTH AFRICA'S 'WEAK SONS'

Notes

1 See for example the following websites:
http://www.hartbeespoortdam.com/
http://www.lesedi.com

http://www.hartbeespoortdam.com/heritage/index.htm
http://www.kormorant.co.za/
http://albums.laurenstravels.com/album15
http://www.myproperty.co.za/town.asp?town name=Hartebeespoort%20Dam
http://www.tcbpublishing.co.za/1Time/abouTimeJul05/hartebeespoortDam.html
http://www.tut.ac.za/tut_web/index.php?struc=1569
http://www.woodstock.co.za/index.php

References

In this chapter 'ACL' refers to Minutes of the various Sessions of the annual Advisory Council of Labour meetings archived in the Central Archives Depot, Department of Labour (ARB). The date of each session is provided, together with the identity of each individual whose statements were recorded verbatim. The sources of quoted documents and statements in this chapter are provided below.

Central Archives Depot (CAD), ARB, Box 5483, Minutes of the Advisory Council of Labour.[complete set]

Government Gazette. 1924. Government Notice 1957. Regulations

Hancock WK. 1968. *Smuts: The Fields of Force, 1919–1950*. Cambridge: Cambridge University Press

Transvaal Indigency Commission. 1908. TP 11

Recommended reading

CAD, ARB. Box 205, LB510/26, Hartebeestpoort Experimental Station

CAD, ARB. Box 209, LB 510/11, Part 3. Complaints. Relief Works. Hartebeestpoort

CAD, Department of Irrigation. Box 36, A89/5, Losperfontein Training Farm: Trainees. Unsuitable Trainees

EPILOGUE

VOICE OF POLITICS, VOICE OF SCIENCE

References

Christie P. 1991. *The Right to Learn*. Second edition. Johannesburg: SACHED/Ravan Press

Dubow S. 1992. Afrikaner Nationalism. *Journal of African History* 33

Enslin P. 1984. The Role of Fundamental Pedagogics in the Formulation of Education Policy in South Africa. In P. Kallaway (ed.) *Apartheid and Education. The Education of Black South Africans*. Johannesburg: Ravan Press

Gould SJ. 1981. *The Mismeasure of Man*. London: Penguin Books

Loubser JA. 1987. *A Critical Review of Racial Theology in South Africa: The Apartheid Bible*. Cape Town: Maskew Miller Longman

Mbeki T. 2005. Speech at the official opening of the Maropeng Visitors Centre at the Cradle of Humankind World Heritage Site, Gauteng, 7 December

NOTES ON CONTRIBUTORS

Marion Bamford is an Associate Professor in the Bernard Price Institute for Palaeontological Research, School of Geosciences, University of the Witwatersrand, Johannesburg. She is a palaeobotanist and her speciality is fossil woods from hominin sites in Africa. She is a Fellow of the Royal Society of South Africa.

Philip Bonner is Professor of History at the University of the Witwatersrand, Johannesburg and holds a National Research Foundation Chair in Social Sciences. He is also head of the multidisciplinary History Workshop.

Jane Carruthers is Associate Professor in the Department of History, University of South Africa, Fellow of the Royal Society of South Africa and Fellow of Clare Hall, Cambridge. She is a well known environmental historian, with a particular research interest in national parks, colonial art and the history of science.

Vincent Carruthers is the director of a management and environmental consulting practice. He has acted as consultant to both government and private organisations in the 'Cradle of Humankind' and assisted with the development of the Institute for Human Evolution at the University of the Witwatersrand, Johannesburg.

Tim Clynick is an independent public policy research executive. He holds a D.Phil in history from Queens University, Canada.

Saul Dubow is Professor of History at Sussex University in England. He is Chair of the Board of the *Journal of Southern African Studies*.

Amanda Esterhuysen is a lecturer in the School of Geography, Archaeology and Environmental Sciences, University of the Witwatersrand, Johannesburg and Director of the Archaeological Resource Development Project, through which she has had a long association with Sterkfontein.

Simon Hall is a senior lecturer in the Department of Archaeology at the University of Cape Town. His current research focus is a multidisciplinary approach to recent Sotho/Tswana history that combines archaeological, oral and written evidence.

Thomas N Huffman is Professor of Archaeology at the University of the Witwatersrand, Johannesburg. He is interested in the archaeology of Pre-colonial farming societies in Southern Africa.

Trefor Jenkins was appointed the first Professor of Human Genetics at the University of the Witwatersrand, Johannesburg in 1975, and has done research into the genetic relationships between living peoples, as well as training and mentoring a large number of clinical geneticists, molecular geneticists and genetic counsellors.

Kevin Kuykendall is a lecturer in Archaeology at the University of Sheffield. His research interests involve field work at Plio-Pleistocene fossil sites in South Africa (Makapansgat Limeworks, Buffalo Cave, Gondolin), and the reconstruction of early hominid life history and patterns of craniodental growth.

David Pearce is a researcher in the Rock Art Research Institute, University of the Witwatersrand, Johannesburg.

Himla Soodyall is a Principal Medical
Scientist/Associate Professor in the Division of
Human Genetics at the National Health
Laboratory Service (NHLS) and University of the
Witwatersrand, Johannesburg.

Goran Štrkalj is a Senior Lecturer in the School of
Anatomical Sciences, University of the
Witwatersrand, Johannesburg.

Phillip V Tobias is Professor Emeritus of Anatomy
and Human Biology at the University of the
Witwatersrand, Johannesburg.

Lyn Wadley is Professor of Archaeology in the
School of Geography, Archaeology and
Environmental Studies, University of the
Witwatersrand, Johannesburg.

ACKNOWLEDGEMENTS AND CREDITS
FOR ARTWORK AND PHOTOGRAPHS

The authors and publishers would like to thank all parties who have contributed artwork and photographs to this publication. A special thank you to Sally Gaule for picture editing the volume, and for her photographs on pages 40, 43, 48, 49, 59, 65, 70 (bottom), 71 (top), 73, 75, 77, 99, 100, 101, 113, 121, 122, 128, 130, 131, 132, 237, 239, and 246.

Particular thanks are given to the following librarians and archivists for their invaluable assistance in sourcing artwork:
Carol Archibald and Kate Abbott of Historical Papers, William Cullen Library, University of the Witwatersrand, Johannesburg
Kathy Brookes and Kenneth Hlungwani at Museum Africa, Johannesburg
Marius Coetzee and Zofia Sulej, Wits Archives; University of the Witwatersrand, Johannesburg
Yvonne Garson for her expertise on the maps of Africa.
Mercy Kgarume, Government Publications, William Cullen Library.
Mmabatho Mackay, Harold Strange Library of African Studies, Johannesburg Public Library
Margaret Northey, Peter Duncan and Fay Blain of William Cullen Library
Francis Thackeray, Heidi Fourie, Tersia Perregil and Stefany Potze, Transvaal Museum, Pretoria

Artwork and photographs on the pages indicated have been provided courtesy of:
Albany Museum, Grahamstown 182
Bell Heritage Trust, University of Cape Town 186 (top)
Brits Town Council Souvenir Album 1924-1974 (1974) 249
Cradle of Humankind World Heritage Site Management Authority, Maropeng 274
Historical Papers, University of the Witwatersrand, Johannesburg 198, 200, 208
MuseumAfrica, Johannesburg 245
National Archives, Zimbabwe 195
Old Mutual Collection (South Africa) 184, 185
Parliament of the Republic of South Africa xviii
School of Anatomical Sciences, University of the Witwatersrand, Johannesburg 30, 31, 32, 34, 37, 72 (below), 73, 74 (above), 76, 77 (top & middle)
School of Geography, Archaeology and Environmental Sciences, University of the Witwatersrand, Johannesburg 17, 102
State Archives, Pretoria 232
Transvaal Museum, Northern Flagship Institution 70 (bottom), 71 (top), 72 (top), 72, (top), 75, 214
William Cullen Library, University of the Witwatersrand, Johannesburg 140, 200
Wits Archives, University of the Witwatersrand, Johannesburg 11, 28, 37, 41, 44, 117, 209, 211

Authors of individual chapters contributed artwork on the pages indicated:
Marion Bamford 92, 94, 95
Amanda Esterhuysen 22, 70 (top), 111, 114, 115, 116, 119, 121
Trefor Jenkins 80, 84
Simon Hall 164, 167, 168, 173
Thomas Huffman 152, 153, 154, 156, 157, 160
David Pearce 136, 137, 138, 139
Phillip V Tobias 217, 220, 221, 223, 226, 227, 230.
Lyn Wadley 133
Brigid Ward is acknowledged for her drawings on pages 149, 154, 157.

Every effort has been made to trace copyright holders of artwork in this publication, but should any oversight have occurred, the publishers apologise and welcome any information that will enable them to amend any errors.

INDEX

This index lists terms, subjects, and personal
names mentioned in the text. Fossil names, site
names, and areas have also been indexed. Figures
are not indexed.

Compiled by Marthina Mössmer

Printed and bound by CPI Group (UK) Ltd, Croydon, CR0 4YY

27/10/2024

14580399-0001